Praise

The Mystery Guest, MG

"What is MG? Can it be cured?"

These questions were the beginning of one patient's odyssey in coping with a difficult illness. It has led her to both old and new medical treatments as well as to explore the possible role of nutrition as it affects her both generally and specifically with myasthenia gravis.

When I first diagnosed her disease, she could barely talk and had great difficulty swallowing. After a thymectomy, medical treatment, and a diet and exercise program, she is able to function almost normally. The author deals with her illness in a logical and unemotional manner. Most important to her recovery has been her faith and determination to get well. …This is a happy story that can serve to help others who face the "Mystery Guest."

> \- Kevin McCoyd, M.D.
> Neurologist, RLT Neurologic Associates, LTD

This book is not just for the patient and family members of those affected by MG. It is a chronicle of faith to meet an unknown affliction and then oneself as the identity of an unwelcome "guest" is revealed. That uninvited guest comes into everyone's life in various forms, sometimes through the body (soma) and sometimes through the mind (psyche). However it comes, each of us digs deep for strength and support to meet it as best we can.

What the author has done in this book is to apply a universally applicable philosophy to a specific disease. Her perspective will serve us all.

- Roger Rydstrom, D.D.S.

The Mystery Guest, MG is a book concerned with nutrition and human health. It meets an important need for persons with a chronic illness, emphasizing proper diet and good nutrition.

I support Peggy's nutritional approach to good health. I admire her faith, patience, and perseverance. And I respect her desire to reach out to others.

- Betty Wedman, M.S.R.D.
  Nutritionist

THE MYSTERY GUEST, MG

(MYASTHENIA GRAVIS)

A PERSONAL PLAN FOR DEALING WITH A CHRONIC ILLNESS

Peggy Matthews Cashman
&
Loretta Kett Bierer, M.S.

Lulu Press, Inc.

N.B. The material in this book is intended to provide information regarding my experience in living with a chronic illness. It is not meant to replace professional medical care, diagnosis, prescription or treatment of any health disorder.

Library of Congress Control Number: 2007909890

ISBN-13: 978-1-4357-0404-6

ISBN-10: 1-4357-0404-5

Cover and interior design by BDDesign LLC.

Printed in the United States of America

TO

JEAN-LOUISE WELCH KEMPTON

This book is dedicated to the late Jean-Louise
Welch Kempton—author, educator,
myasthenic, mentor, and friend.
She was an inspiration to me and countless
others. PMC

To our husbands Jack and Bion for their love,
support, and encouragement.

CONTENTS

CHAPTER 4
THE SUBPLOT OF ACCEPTANCE 61

CHAPTER 5
SEARCHING FOR CLUES 82

CHAPTER 6
THE PATH OF NUTRITION 107

PART II:
ANALYZING THE MYSTERY

INTRODUCTION

When I was a young girl the letters "MG" meant a snappy little sports car. I've learned, however, that these letters represent a rare disease called myasthenia gravis, a combination of Greek and Latin words meaning serious muscle weakness. Almost thirty years ago, I was diagnosed with myasthenia gravis (MG). My illness was identified after a long struggle with baffling symptoms. Because of this difficulty, I also think of the letters—MG—in terms of an abbreviation for "mystery guest."

You may recall (if you are a person of a certain age) a popular TV game show called *What's My Line?* A celebrity panel had to guess the occupation of guests by asking a series of questions. Often the responses and the queries were extremely funny. At the end of the show, the panel would don their eye masks, similar to the one on the book cover, while a well-known person would "sign in" on the blackboard. The panel would then proceed to question the mystery guest in an attempt to identify the famous individual. In my case, the uninvited mystery guest of MG came to visit me and it appears to have come for a long stay.

The theme of the unknown permeates *The Mystery Guest, MG (Myasthenia Gravis).* In Part I, Resolving the Mystery, I recount the story of my journey with a chronic illness—the initial symptoms, difficulty in finding a diagnosis, and the mounting pressure as the signs of muscle weakness progressively worsen. Once MG is identified and confronted, I detail my research to discover ways to boost my body, mind, and spirit. The fruits of this pursuit result in my personal plan for dealing with a chronic illness.

Part II, Analyzing the Mystery, provides a simplified layperson's explanation of how MG works and the tests, treatments, and therapies currently available. It also highlights medical and community resources, particularly internet web sites. This section concludes with the latest medical research findings which offer abundant hope for the future.

Although this book is filled with pertinent facts and research, I have gone beyond information to bring you a story of one person's ordeal filled with loving, caring, suffering, and surviving. I don't purport to give medical advice nor do I suggest a cure for MG or any other illness. Everyone is different in biochemistry as in appearance, and what is beneficial for one person may not be good for another. I encourage you to always rely on medical professionals to guide you in health issues.

Health is something most of us take for granted while we are well and enjoying life. Often it is not until we lose our health that we try to regain it. Sometimes our attempt to recapture wellness is too late. This book is written for all those who suffer from a chronic illness, their caregivers, family members, and friends.

My faith strengthens and sustains me. I believe the Lord gives all of us "mountains to climb" and the resources necessary to scale those heights. Even though challenges may continue unabated, our perception can change through faith and prayer. How we respond to human suffering is a familiar theme. No one is immune to life's trials and defeats; we all have our disasters and disappointments. How we handle our difficulties depends on our attitude which can be positive or negative. You and you alone must choose which attitude you will have. Fortunately, we don't have to face our problems alone. Our belief system in a God greater than ourselves can give us insight, guidance, comfort, and strength when we're most desperately in need.

If you are faced with an uninvited guest, a mystery illness of any nature, I encourage you to meet it with a positive attitude. Set goals for yourself and trust that your condition will improve. This striving can be a burden requiring patience and perseverance. But if you accept the challenge and persist, you may discover that you have grown in compassion, strength, and inner peace. It can be your greatest hour.

PART I

RESOLVING THE MYSTERY

CHAPTER 1

A MYSTERY GUEST ARRIVES

*T*he snow was falling steadily and creating drifts in the street below. I thought of how each six-pointed snowflake is different in size and shape. While it is not unusual to have a major snowfall in Chicago during the month of December, the beauty of it is not always fully appreciated. As the temperature plummeted, ice formed on the trees and I almost felt transported to a scene in Dr. Zhivago.*

I was in Foster McGaw Hospital of Loyola Medical Center for a Tensilon test to determine if I had myasthenia gravis, a rare autoimmune disease. My window overlooked a large woodland area with a highway winding between the forest and the hospital. I could see the busy world of automobiles with people rushing here and there, and the quiet contrast of the peaceful woods. It was the Christmas season and there were colorful decorations throughout the hospital. I could hear carolers in the distance.

Suddenly the door opened and my physician entered with a team of doctors and nurses. Loyola Medical Center was (and is) a teaching hospital and I was there for a test rarely performed and observed. My husband Jack, and my roommate Sue and her mother (who had heard nothing but my slurred speech) were also present. The anticipation and suspense surrounding the test were remarkable because the results would answer so much uncertainty. I would know if I had myasthenia gravis, a rare autoimmune disease in which each myasthenic—like each snowflake—is unique. The test would be the first step in solving a medical mystery that had defied diagnosis. I wondered how this mystery guest would change my world.

EARLY SIGNS

Strange things were happening. I had no pain but I knew something was wrong. When I woke up in the morning I felt great. By mid-morning my face would appear "sleepy" and my smile became a frown. The muscles around my eyes also began to weaken. I had a hard time keeping my eyelids open, yet during sleep I could not always completely close them. This exposure caused my eyes to dry out during the night. When I awoke my eyes were sore and burning.

My speech began to slur and was nasal in tone. Frequently, I would have to clear my throat and I often had a hacking cough. As a child I had bronchitis, so I recognized some of these signs as weakness in my respiratory muscles.

By the end of each day I had little control of the muscles in my throat, mouth, and eyes. My speech, which was clear for a few words, became nasal and then slurred as the palate muscles in my mouth weakened. Swallowing was difficult. At the beginning of a meal, I could chew a piece of steak normally. By the end of the meal, the food often dribbled out the sides of my mouth. All of these changes made me feel "out of control." I thought it was odd that my muscle weakness increased with use while my strength was restored with rest. In addition, all of these unsettling symptoms were not constant; they varied considerably from time to time.

A FULL HOME LIFE

My illness came as an unwelcome surprise at a time in my life when I was going through many transitions. We were a busy family, often too occupied to notice the change of colors in the fall or to count our many blessings. Our family of four lived in a comfortable suburb of Chicago where my husband, Jack, was General Counsel for the Midwest territory of a large corporation. While I had previously applied my business and secretarial skills working for an oil company in Texas, I was now devoting my energies to my family and community services. We entertained frequently and I took pride in being a good wife and mother. I was also completing my studies in interior design at a local college with plans to start my own business. I was timing this new endeavor to coincide with the departures of my son and daughter for college. A business in interior design would give me a chance to be creative yet allow time to entertain and travel with my husband.

Suddenly I found that I was no longer in control of my carefully planned life. My increasingly frequent bouts with an unknown illness were beginning to dominate my day-to-day existence. I knew I had to find out what was happening to me.

On a personal note, I owe my sanity to my family. I draw great strength from them. My husband, Jack, concentrated all of his energies toward helping me during my time of need. My children were also very supportive. When my problems first surfaced, my son John was a junior in high school. He lettered in soccer that year, was elected to the National Honor Society and named an Illinois State Scholar. During the summer he taught swimming and lifesaving at the Park District pool where he was a lifeguard. It was also the year John earned Eagle Scout, the highest rank in the Boy Scouts.

My daughter Susan was a freshman in high school. She was active in Young Life Crusades for Christ and taught Sunday School. Susan was an honor roll student and an active member of the high school gymnastics team. During the summer, she was a member of the "Y" gymnastics team which achieved first place at the State Gymnastics Meet. It was not unusual for Susan to enter a room with a flip-flop or a somersault. She was exhausting to watch but a pleasure to have around. Each day after school, I looked forward to her cheerful chatter and the more serious concern from my son.

4

Little did we realize during this happy and hectic time that our lives were about to change radically. With the encouragement of my family's support, love, and understanding, my illness became an ongoing challenge and a source of growth for all of us. We have learned how important it is to "take time to smell the roses."

A GROWING CONCERN

It was the Bicentennial year 1976. Spring came but I did not have my usual energy. The previous year had been full of activity and frustration at my constant feelings of fatigue. I was also nervous and impatient and, from time to time, my speech would slur. A checkup with my internist led to several tests including an upper GI exam which revealed no major health problems. A friend suggested it might be the "change of life." At the age of forty-three, it was certainly possible but I sensed that it was more than that. Still there were times when I wondered if perhaps I was exaggerating my difficulties, especially when there were instances when I was perfectly normal.

This relapsing pattern continued and then, in June 1977, a series of eye problems began to trouble me. I had swelling around my eyes and my eyelids were droopy. Another friend half seriously suggested that my droopy eyelids might be a sign of "advancing" age. While I appreciated her reality check, I knew that my eye problems were more complicated than the gravitational "pull" of maturity. Rubbing my eyes

5

triggered acute conjunctivitis which was treated as a possible allergy by an ophthalmologist. The doctor was puzzled by the weakness of my eyelids. He observed that my eyes would not completely close—which caused drying—and, at the same time, it was difficult to open my eyes wide because of the drooping eyelids.

The ophthalmologist suggested that I consult a neurologist to find out why my eyes would not close. In hindsight, how I wish I had taken this early advice and followed his suggestion. But I was also seeing an internist and I had every confidence that he would find the solution. Surely it was only a temporary or minor inconvenience that would resolve with patience.

During this time of searching, my dermatologist recommended that I stop wearing makeup—not an easy task at my age but I was desperate and so I tried it. The swelling subsided but the muscles around my eyes still did not feel normal. I remember trying to explain it by saying, "I don't seem to have control of the muscles around my eyes." Later this weakness spread to my mouth and tongue, a phenomenon referred to as a "sleepy face." Then the muscles in my neck began to tire easily. Frequently I found myself using my hands to hold my head upright or leaning against a pillow or backrest to support my neck.

By December, I actually felt somewhat better or perhaps I was getting conditioned to the on-again/off-again symptoms. Our family decided to celebrate by having a Christmas Open House to add to the festive season. We had always taken great pleasure in entertaining our friends. Preparing for a festive party, especially a theme-based celebration, was one of my favorite things to do. The Christmas holidays made it simple. I would carefully plan my menu, cook most things ahead of time and order in for the remainder. Music was an important part of the day and we had help with the serving. I was feeling great when our first guest arrived. By four o'clock that day, my eyes felt as if they were in spasm and would pop out of my head. I lost control of the muscles around my eyes and mouth, making me incapable of even saying goodbye to many of our guests.

We cancelled invitations to several Christmas parties that year. I would get ready to go and then realize that I couldn't handle conversation with my fading voice. My eye muscles were so unsteady that I would blink constantly. And the muscles in my neck were so weak that it was hard to sustain my head in an upright position for a long period of time. My husband was very solicitous and perhaps a little frightened. It was very unlike me to *not* want to go to a social gathering. Something had to be seriously wrong.

I have since learned that while weakness is a symptom of many diseases, the physician should suspect MG when the patient complains of weakness that involves specific muscle groups, especially the ocular (eye) and oropharyngeal (upper part of mouth and throat) muscles. Slurred speech, difficulty chewing or swallowing, and muscle weakness in the neck are also common.

Other warning signs of MG are: heaviness in limbs or abnormal fatigue of the muscles (not tiredness) especially after repeating an activity; ptosis (droopy eyelids), and an expressionless facial appearance due to weakness of facial muscles.

I also learned many factors can worsen myasthenia. A woman's menstrual cycle can exacerbate muscle weakness. Hot weather or a hot bath, infection of any kind, thyroid dysfunction, illness, emotional upset, drugs, and surgery may also cause increased weakness. I experienced most of these fluctuations before I was diagnosed.

TESTS...AND MORE TESTS

If you are reading this book, chances are that you are dealing with a chronic illness, either directly as a patient or indirectly as a caregiver. I have no doubt that the following pages may have a familiar ring.

I began a long series of monthly visits with an internist. He took numerous blood tests as well as X-rays. A lump in my throat proved to be a small growth that did not touch the thyroid. I was told not to worry about it. A thyroid scan and other tests did show that I had a slightly underactive thyroid. The doctor advised me to start taking one milligram of Synthroid which is a synthetic thyroid hormone supplement. Over a period of several months, he gradually increased the dose.

By the summer of 1978, I was feeling better. I had more energy but I was very nervous. I still had the droopy eyelids. If I became overly tired, I had difficulty controlling the muscles around my mouth and eyes. Eating was complicated and messy. When we went out in the evening, we usually returned home by nine or ten o'clock because I was exhausted. I would feel fine one minute and then collapse like a car that had "run out of gas," a common sensation in myasthenics.

When I saw the internist again in September, I was convinced something was definitely wrong with me. A year had gone by and I had learned nothing new. I told the doctor that I had no pain but I was tired, very nervous, and easily disconcerted by minor difficulties. I repeated my description of the weakness in my face and neck muscles and stressed that I was not getting better, I was getting worse.

I also described the nasal tone in my voice and the fact that my speech was slurred, especially at the end of the day. (My good friend Nancy thought I was drinking alcoholic beverages because of my slurred speech. She very graciously confronted me about the matter and I assured her I was not inebriated. In fact, I was in the process of consulting with an internist to determine what was wrong.)

Swallowing both liquids and solids gradually became a major problem. I found myself drinking milkshakes and soup and, when even these became difficult to swallow, I was frightened.

The internist had ruled out most of the major internal illnesses. He indicated that my problem was strictly an emotional one and that I was not telling him everything. He was convinced that my problem was psychosomatic! Even though I assured the doctor that my husband was not seeing another woman and my two teenagers were not causing problems, it was obvious he did not believe me.

At this point, you may wonder why I continued seeing a physician who obviously was not convinced that my symptoms were genuine. In retrospect, I find it difficult to explain my own viewpoint and emotions at the time.

Those of you who struggle with a chronic illness will probably understand that I was physically drained much of the time. This lack of energy and the stress of dealing with baffling symptoms were not conducive to making decisions or thinking clearly. The internist was a well-respected professional and he seemed to reach his conclusions after a lengthy period of testing and searching. In addition, I was truly bewildered by my on again/off again symptoms and easy prey to questioning myself.

MORE DISTRESSING DEVELOPMENTS

Later that September my husband and I went to Florida. My eyelids were heavy and tired, and I needed to go to bed early each night of our vacation. I began to avoid conversation, contrary to my usual gregarious nature, because of the weakness in my throat muscles. My feeble facial muscles flattened the expression in my face, although this change in appearance was barely noticeable to most of the people around me.

During the first few weeks of October, I seemed to reach a plateau. Even though I did not feel normal, I was able to play tennis once a week and "appear" to be alright. I was still drinking lots of milkshakes and gradually realized that my difficulty in swallowing was becoming more frequent. Then suddenly there were some new developments. I began to break out in heavy perspiration, a phenomenon that would last about five minutes and then reoccur almost every hour. At the same time I experienced a rapid heart rate. What did these signs mean? I knew the heavy perspiration could be any number of things. I could indeed be in the throes of the "change of life" as several of my friends had suggested. It could be a typical signal of respiratory insufficiency. Heart failure or hyperthyroidism can also trigger rapid heart rate and heavy perspiration. Even medication, such as the Synthroid I was taking, can cause these symptoms.

Once again I was alarmed by the increasing evidence that something was definitely out-of-order. My calendar was full and I was trying to keep all my engagements in the false hope that I could stay busy enough to forget my problems. Incredibly, my internist remained convinced that my difficulties were of a psychological nature and would subside in time. And I continued to trust his judgment. Perhaps I did not want to believe that I had something seriously wrong. I clung to false hopes and denial.

On November 2nd, I attended a church luncheon with friends. I could barely eat and was breaking out in perspiration. The nasal tone in my voice was very evident and my speech was slurred. Overall, I felt very weak and out-of-control. I left the luncheon early and went to the doctor's office and asked to see the doctor, although I could hardly speak clearly enough to be understood. After trying to explain myself, the doctor asked—once again—what was wrong at home. This response aggravated my speech even more and I soon left his office more upset and discouraged than ever.

My droopy eyelids, slurred speech, nervousness, perspiration, and fatigue persisted for the rest of the month. I was used to being active and determined to continue to participate. At the time I was not aware that my body was desperately in need of rest. I even managed to play doubles' tennis. Blurred vision and excessive blinking caused me to completely whiff the ball occasionally, but somehow I always managed to complete the hour-long game.

However, my slurred speech and difficulty swallowing were impossible to overcome. My husband and I began to cancel weekend commitments. I realized that I had to get my health problems resolved.

SEARCHING FOR ANSWERS

On Thursday, November 16th, I saw the internist again and heard the same story, "What is wrong at home?" This was the final blow. At last, I told the doctor that I needed to find answers and asked for a referral to another specialist.

Cancer had always been a concern for me. My father had died of cancer of the lungs and stomach, and several of his family members had also died of cancer. My internist was also a specialist in cancer research and had ruled out cancer in my case. His referral to an ear, nose, and throat specialist confirmed my cancer-free status. The ENT doctor said, "I don't know what you have, but you do not have cancer of the throat or the larynx." Feeling somewhat relieved, I muddled through the next week and had dinner guests for the Thanksgiving feast—a pleasant but tiring day.

Monday, November 27th was a snowy and blustery day. The roads were icy and my husband suggested that I skip my early morning interior design class. At first I thought I would follow his advice, but I didn't want to pamper myself. Instead I dressed warmly and left for my eight o'clock class. The roads were slick but I reached the school with no difficulty.

Instead of going to the classroom, I proceeded directly to the professor's office and asked to speak with him. I tried to explain, "I have some problems and I might have to miss school for awhile." (This meeting was completely unplanned and I had no idea I would be in the hospital for tests within two days.)

The professor asked if we could talk after class. I just looked at him, thinking I really had nothing more to add. Then he turned to me and said, "To be honest with you, Peggy, I didn't understand a word you said." I realized I had been mumbling and slurring my words again. To put him at ease, I laughed and scribbled my message to him on a sheet of paper. He smiled and promised not to call on me during the class. Somehow I survived the two hour lecture and headed for home.

On the way, I decided to stop at a nearby shopping center and do some Christmas shopping. In a few hours I accomplished more than I would normally have done in two months. I sensed that the Lord was in control of my life that day in guiding me to alert my teacher and helping me to prepare for an important family celebration.

By one o'clock I was home but I could not eat my lunch. My symptoms seemed to be worse than ever before. I decided to go to my doctor's office so that he could see my deterioration firsthand. At this point, my mumbling speech was unintelligible and I had to explain my situation by writing notes. Because he was alarmed at my obvious decline, my doctor scheduled an appointment for me to see a neurologist the next day.

At five o'clock the next morning, I woke up wondering how I would communicate with a new physician when I couldn't speak. Sleep was out of the question. I got out of bed and started typing all that had happened to me for the past two years. Once again, I knew that the Lord was leading and preparing me, and helping me to find the words.

After attending a funeral for a friend's father, I drove to my appointment with the neurology specialist. The streets were icy and it was snowing. At first I attempted to describe my symptoms but my slurred speech made that impossible. Instead I gave the typed account to the neurologist for his information. Finally, after years of confusing and mysterious symptoms, I was about to discover some answers.

****

Within three minutes after the neurologist read my summary of symptoms, he gazed at me intently and said, "I think you have myasthenia gravis."

It was a bittersweet blow. On the one hand, I finally had a name for all my various and bewildering symptoms; on the other hand, I was in unknown and frightening territory.

I asked, "What is myasthenia gravis and can you cure it?" He told me that myasthenia gravis (MG) is an autoimmune, neuromuscular disease that means severe or grave muscle weakness. Even though it is called neuromuscular, MG is not a disease of either the nerves or the muscles. It is a disorder in the gap which lies between nerves and muscles. In a normal nerve impulse to a muscle, the brain stimulates nerve endings to release a message-conducting chemical. This chemical messenger is called acetylcholine which is abbreviated ACh. By transmitting the message from nerve endings to receptor sites on the muscle, ACh stimulates muscle contraction.

Generally, there are many receptor sites in the gap called the neuromuscular junction. In people with myasthenia gravis, however, the number of receptor sites is greatly reduced. Because the "chemical message" from the nerve to the muscle is blocked, the normal contraction of the muscle is incomplete and difficult to maintain. In myasthenics, this incomplete transmission of nerve-to-muscle message results in weakness that worsens with each contraction.

This helped me understand why I was fairly good in the early morning but much worse later in the day. It also explained why my symptoms of weakness were aggravated by activity and improved with rest.

SYMPTOMS OF MG

As I tried to absorb the neurologist's explanation, I wondered about my specific symptoms and how they fit into this picture. I learned that symptoms in most patients were milder upon rising in the morning and grew worse as the day proceeds. Loss of facial expression is often seen. Impaired throat muscles cause unintelligible, slurred speech with a distinctly nasal quality imparted to the voice. The involvement of the mouth, throat, and food tube muscles may cause nasal regurgitation of fluids, choking, and aspiration of food, very unpleasant happenings. A weak cough sometimes develops from depleted respiratory muscles. This can lead to an inability to clear the respiratory (tracheal and bronchial) tubes of mucus and fluids. Weak respiratory muscles may also cause breathing difficulties. In some myasthenics, the muscles in the arms and legs are the most affected. Some patients cannot walk and can hardly sit up. Others have trouble lifting their arms. I could identify with many of these problems.

While the doctor described common symptoms of MG, he emphasized that the kinds and severity of symptoms differ considerably from patient-to-patient. Each myathenic usually experiences varying degrees of symptoms.

Now the disconcerting nature of my problems, the on-again/off-again symptoms made more sense. No wonder I felt so good when I awoke in the morning and then could barely eat or speak by the end of the day. The knowledge and compassionate understanding of the neurologist filled me with gratitude. At long last I was beginning to find answers.

Of course, the neurologist needed to confirm his preliminary diagnosis. And I had to confront life with MG in earnest with the realization that this uninvited visitor was here to stay.

(You'll find a more detailed explanation of MG in Part II, Chapter 9, Investigating How MG Works.)

POSTSCRIPT

As I look back on that snowy evening while I waited for confirmation of the diagnosis, I realize I had no idea of the challenges ahead. Perhaps it is a blessing not to "see" into the future. One day at a time, one step at a time...

I am reminded of all the things I am privileged to enjoy despite my MG and some additional health challenges. Even though my vagus nerve was damaged during brain surgery in 2001, I can taste the spicy tang of my husband's tortilla soup and the creamy richness of my favorite dessert, crème brûlée. I appreciate that I can walk despite persistent pain. Without setbacks and difficulties, it is easy to take the good for granted. Now I take time to "smell the roses."

CHAPTER 2

THE PLOT THICKENS

Before I continue with my story, let's explore what happens when you hear the words, "You have____." What are your first thoughts? This expression is frightening and sometimes devastating, regardless of what disease you may have or how strong you are. There is a sense of desolation which can be overwhelming. Tragedy occurs, injustices exist, and in the words of Rabbi Harold Kushner, "Bad things happen to good people." Where does a person turn for help? If you or someone you love has MG or any chronic illness, you will probably consult more doctors and undergo more tests in a year than many people experience in a lifetime.

When I went to the neurologist for my initial visit, it was the first time I remembered hearing of myasthenia gravis. As I was trying to absorb the reality of finally identifying my illness, I was also told that I should immediately enter the hospital. Specialized tests were needed to confirm the diagnosis.

The neurologist offered to call my husband and explain my situation. I can laugh about it now, but at the time I seriously told the doctor that he couldn't bother my husband. Jack was joining friends after work to have dinner and see the opera "Electra." While I usually loved going to the opera, I didn't care to attend the performance of this heavy German production. At the same time, I did not want to upset my husband's plans.

The doctor didn't know me or my husband, of course, and he just looked at me in disbelief. I knew that Jack had looked forward to this opera for months. To placate the doctor, I promised to tell Jack the following day. (Later I learned that the doctor was concerned that I would not get through the night without the disease affecting my breathing. Because of the potential for a respiratory crisis, the doctor urged immediate hospitalization.) After some negotiating, the neurologist persisted and I somewhat reluctantly gave the doctor my husband's business card with his telephone number.

The neurologist was direct and blunt. He told my husband, "I believe your wife has myasthenia gravis. I think it is imperative that she be checked into the hospital immediately because I need to confirm my diagnosis."

Jack asked a number of questions and then made a fast exit from his office, reaching our home about the same time I did. After much discussion, I had prevailed upon the doctor to give me a day to prepare myself and my family. With some hesitation he agreed to the postponement with the proviso that I call him if there were further complications.

Those who know my organizational nature understand my dilemma. The laundry had to be up-to-date and the house tidied before I could possibly check into the hospital. Again, my concerns seem ludicrous in hindsight. More significantly, I needed time with my husband and children, and the chance to ask friends concerning which hospital and medical staff would be best for me.

PREPARATION TIME

During my brief reprieve, we researched our area's medical facilities. We discovered that Northwestern University Medical School in Chicago, the University of Chicago Medical School, and Foster G. McGaw Medical School at Loyola University in Maywood, Illinois were all noted for their work in myasthenia gravis. By going to Loyola, I could continue with the neurologist who had finally identified my "mystery guest." It was a much needed and appreciated comfort. This physician was not only fairly certain of his diagnosis; he also had a plan to help me.

While I was home preparing to enter the hospital, I recalled reading in the newspaper about Aristotle Onassis having had myasthenia gravis. Actually, that celebrity news item was probably the first time I had ever heard of MG. During the emotional revelation in the doctor's office, I had not connected my diagnosis and the earlier news.

As I waited for the confirmation of the doctor's diagnosis, my thoughts were spinning in all directions. The distress of knowing that something was wrong yet uncertain what lies ahead was unsettling. Now I was waiting again for tests to confirm the diagnosis and the possibility of major surgery. You may have experienced these emotions. Nobody said life is easy, trouble-free or without problems. Waiting for the answers is the hard part.

I realized that it is possible to experience fear and courage at the same time. Fear tells us life is unpredictable, anything can happen. At the same time courage replies quietly, "God is in control." Yes, there were times when I wanted to be in charge. But without God, life has no purpose and without purpose life has no meaning. And without meaning, life is without hope to sustain it. So in the turmoil and confusion of all these conflicting feelings, I chose to place my trust in faith, recognizing the fear but forging ahead with courage.

IN RETROSPECT

A probable diagnosis for all my varied symptoms also stirred me to think of when the first signs of trouble occurred. In the summer of 1974, I joined my husband on a business trip to the Lodge of the Four Seasons in the Missouri Ozarks. While Jack attended meetings, I spent most of the day in the sun. It was my first time in the sun that year and I got badly sunburned. To add to the overexposure to the sun and its unpleasant aftermath, I also skipped lunch that day. Then we attended a cocktail party that evening honoring the guest speaker who was the general counsel of a large corporation in the Chicago area.

After the speech I started to walk across the patio to greet the speaker. Suddenly my legs gave way and I collapsed. Since I hadn't had a drink, I thought perhaps I was experiencing the aftereffects of sunstroke or heat exhaustion. Jack helped me to my room where I ordered dinner and rested for the remainder of the evening.

The next day I was fine and had no ill effects. Since then, I have learned that extreme heat worsens the muscle weakness and fatigue of myasthenics. So it seems probable that I had the first symptoms of MG as early as 1974.

On another occasion that same year, I was in my kitchen preparing dinner. In the midst of chopping vegetables, my right hand became limp and I could not hold the knife. Fortunately, my daughter came to my rescue. While this muscle weakness was temporary, it was another distress signal.

There were other early warning signs. For example, sharing a meal at a gathering with friends sometimes precipitated difficulties. During one dinner, my speech became slurred and I broke out in perspiration. My dear friend Anne, who was aware of my problems, suggested that perhaps we should leave early. I remember being upset and angry that I had spoiled our evening.

On another occasion, my husband and I were invited to our friends' home for dinner. While I gave the hostess a hand in the kitchen, she asked me the simple question, "How are you?" When I attempted to reply, my speech was so garbled that I burst into tears. Of course, she thought something was terribly wrong. Jack explained that I was having all sorts of unusual symptoms which had not yet been identified by our doctor.

Later, when I tried to eat, food ran out the sides of my mouth. My friend, Delores, had gone to considerable effort to prepare a beautiful gourmet meal, but my attempt to eat failed miserably. I tried to laugh about it, but I felt terrible that I had ruined an otherwise lovely evening. ...Now all of these random happenings took on a deeper meaning. They fit a pattern that could be identified as *myasthenia gravis*.

HOSPITAL TESTING

My thoughts were racing as I entered Foster McGaw Hospital at Loyola University Medical Center for tests to confirm the MG diagnosis. I realized there were many options in the treatment of myasthenia gravis. It was reassuring to know that my doctors were knowledgeable about the disease and had successfully treated MG patients. And, finally, I would find out which direction my care would take.

There is nothing comfortable about entering a hospital. By the time all the admitting procedures were completed and I was finally in my room, I was so exhausted I could barely speak. My husband, Jack, was tense and nervous. And I was anxious to begin the testing necessary to definitely confirm the diagnosis.

Just as I was settling into my hospital room, the door opened and a young, attractive girl walked in with her mother and sister. Her name was Sue and she was to be my roommate. Sue was in the hospital for tests, also. Jack was relieved to have someone to talk to and it made me calmer to see him relax. As they chatted away and Jack described my difficulties, I focused on the thought that soon the tests would be over and some resolution would be reached.

There are three tests that are used to confirm the diagnosis of MG: an injection of a muscle booster drug, an electrical stimulation of the neuromuscular system, and a blood test that measures the level of antibodies to acetylcholine receptor. During this hospital visit, I was given the first two tests.

Edrophonium chloride or Tensilon is a short-acting muscle stimulant that the doctors inject into the vein and then observe the patient's response. This test was developed by Dr. Kermit Osserman at Mt. Sinai Hospital in New York City in 1952. The procedure causes a startling change in symptoms when the results are positive. Certainly it left an indelible impression on my mind.

The doctors entered my room and, after checking my identity, they called all the nurses on the fifth floor (neurology) to observe the test. Although they were well-informed about MG, this was a teaching hospital and some of the staff had not witnessed a positive Tensilon test. I sat up surrounded by my husband and at least a dozen strangers watching motionless as if I were some unusual specimen in a side show. They waited eagerly to see what change—if any—would occur.

The injection was made. Like magic, my masked expression melted away and my eyelids lifted. And wonder of wonders, my voice was clear and strong. In addition to giving thanks for this reprieve, I could also tell them I was hungry. The nurses scurried to bring me food, knowing that this "miracle" was not a lasting change.

I'll always remember the look of complete amazement on the faces of all those present. I actually came "alive" for about four minutes. This has to be one of the most dramatic reversals experienced in medicine—weak muscles being restored to full function. And then, just as quickly as my symptoms had subsided, they reappeared. The masked expression as well as the droopy eyelids returned. My voice reverted to mumbling sounds. It was over. The test was positive for myasthenia gravis. As difficult as it was, I sensed relief to finally know what was wrong.

The second test of stimulating nerve-muscle interaction is used primarily to determine the extent of the disease. It also helps to differentiate MG from other neuromuscular diseases. The electromyography test involves stimulating a peripheral nerve by means of an electrode placed over the surface of the nerve. In my case, the electrode was placed over the ulnar nerve in my left wrist and electrical current was then applied to that nerve. Stimulating the nerve causes the muscle in the hand to contract. Electrical activity in the hand muscle is then recorded on a surface electrode, pumped through an amplifier and displayed on a TV monitor.

My doctors told me this test would be painful but that it only lasted twenty seconds. Prior to my entering the test room, I had heard screams coming from (what turned out to be) a robust, young man. I smiled and said confidently, "I can take anything for twenty seconds."

I must confess, however, that my smile faded quickly and I probably screamed louder than the young man before me. To make matters worse, the technician conducting the test ran out of tape fourteen seconds into the test. Yes, the complete test had to be repeated. The electromyographic test was also positive for MG, confirming that the disease was generalized.

(In 1995 and again in 2006, I had this same test which is now greatly refined. While the procedure is not pleasant, it is a "piece of cake" in comparison to my first experience. I am amazed at how the medical field has improved over the last two decades.)

THE SPECIFIC DIAGNOSIS

The clinical name for my condition is *Ptosis Dysarthria Severe Generalized Myasthenia Gravis.* Translated this means:

ptosis—the drooping of one or both eyelids;
dysarthria—involving muscles of speech,
chewing, and swallowing;
generalized—affecting all of the body.

While I was grateful to finally have a definite diagnosis, it was frightening to learn that I was in the late severe category in which the estimated response to drug therapy and prognosis is poor. In addition, it was extremely distressing to realize that had my internist referred me earlier to a neurologist, I might have been diagnosed and treated before reaching this late, severe stage.

SURGICAL APPROACHES

My doctors advised an immediate thymectomy, removal of the thymus gland, by the transsternal breastbone-splitting approach. It would also require the surgeons to perform a tracheotomy, a surgical opening in the trachea or windpipe, as a preventive measure in case breathing problems occurred during the surgery.

While surgery to remove the thymus gland was not new in 1978, very few doctors were using it for the treatment of MG. Since my neurologist diagnosed me immediately and was familiar with the procedure and its potential benefits, I was confident and comfortable with his recommendation.

(Today, approximately 75% of myasthenics have some thymic abnormality and 15% have a tumor of the thymus gland which is called a thymoma. Therefore, many doctors use thymectomy for at least some of their MG patients. The current research on thymectomy is discussed in Chapter 12.)

I asked, "How will surgical removal of my thymus gland help me?" I was told that patients with generalized myasthenia gravis usually benefit from a thymectomy. The symptoms of MG lessen and there may even be a remission in some cases. If there is no tumor involved, the prognosis is generally brighter. Patients with thymomas also benefit from thymectomy because of the danger of tumors invading other structures in the chest.

While the doctors could not give me total reassurance, I was heartened by the fact that two-thirds of myasthenics benefit significantly. Although advances of the past two decades have essentially eliminated the risk to life, thymectomy is a major operation. I was advised that it should only be undertaken with a thorough understanding of what may or may not be accomplished.

I should mention that there is another operative approach in surgically removing the thymus gland. In this method the incision is made above the breastbone at the base of the front of the neck. Although this transcervical approach is simpler and causes less discomfort, it is not suitable for removing thymus tumors of significant size. Because this procedure offers poorer visualization of the surrounding area where the thymus may be located, total removal is somewhat less likely.

It is important to keep in mind that thymectomy (by either method) is an elective and not an emergency surgical procedure. However, this operation seems to offer the best chance of any form of treatment for lasting improvement or remission.

At the present time, thymectomy is often the preferred medical treatment for adult patients with generalized myasthenia gravis. Thymectomy should only be performed in medical facilities thoroughly familiar with the necessary operative and postoperative management of patients with MG.

Most surgeons favor a sternal-splitting approach to thymus surgery which permits the most complete removal of thymic tissue. Thymectomy is generally not performed in patients with ocular myasthenia, although there is some evidence that it may be beneficial.

THE NEED FOR SURGERY

I agreed to have the transsternal thymectomy my doctor recommended. It was Friday and the surgery was scheduled for early Monday morning. During this time, I needed to have blood tests and get plenty of rest while the nurses observed my attempts at eating. I was feeling good; I just couldn't speak and swallowing food was difficult.

My roommate was young, energetic, and cheerful. We both had so much pent-up emotion that we decided to reach out to brighten our hospital surroundings. It was the Christmas season—the halls were being decorated and carolers could be heard in the distance. In the spirit of the season, we decided to join in the festivities and decorate our room. With green tape from the nurses, we fashioned a Christmas tree shape on the wall. Then we taped greeting cards to the tree and put other cards on the window. Our room overlooked a forest preserve. The snow was heavy and deep, making our view spectacular.

Plants and flowers began to arrive. Friends from all over the world—Europe, Washington, DC, Texas, California, and the Chicago area—sent flowers. It was heartwarming to know so many were thinking of me, but a little embarrassing when my roommate had none. Fortunately, my thoughtful husband sent flowers to my roommate. Sue was thrilled and I enjoyed my "garden" even more.

After my surgery on Monday, I knew I would be in the Intensive Care Unit for at least a week. Since plants and flowers were not allowed in the Intensive Care Unit, I sent some of them home on Sunday. Then Sue and I spread joy on the fifth floor, sharing the beautiful flowers with other patients. It made us happy to see their surprise and it helped to expend some of our own nervous energy.

Finally, Sunday night came and I was sedated to ensure a good night's rest before the operation. The sedative medication seemed to have the reverse effect on me. I could not sleep. I turned on the radio and thought about my children and my husband. How much I missed them. I thought about the love my husband Jack and I shared. We were (and are) independent persons with a strong bond of common values. I reflected on all our good times together over the years.

It was after midnight and I was listening to soft, spiritual music on the radio. As I lay awake, I trembled at the thought of major surgery. I prayed, "Lord, I place myself and all my weakness in Your hands. Please help me to face this surgery with courage. And guide my surgeon and the medical staff during the operation."

I knew that at my age my thymus gland should be dormant, but major chest surgery and its aftermath frightened me. The minutes ticked by and I continued to pray until I finally fell into a deep sleep.

When I awakened the next morning, I heard voices. I was greeted by my husband and his good friend, Roger. Roger had taken the day off from his dental practice to be with Jack. I was grateful for Roger's support because I had images of Jack pacing the floor alone while I was in surgery. Now Jack could walk the floor with a friend who would help sustain him.

Certainly, the presence of Jack and Roger was a big boost to my morale. As I went into surgery, I realized that others faced ordeals far worse than mine. Oddly enough, this thought also helped to bolster my courage.

My next visitor was the anesthetist who gave me a sedative injection to help me relax. Then a young orderly and a nurse came to roll me to the surgical room. The five of us crowded into the elevator and started down to the operating floor.

By now, I was feeling calm and lightheaded from the sedative. But there was an apprehensive quiet in the elevator. I tried to think of something humorous to break the deadly quiet as the elevator descended to the surgical arena. Then the nurse asked me about my personal belongings. I replied, "I gave them to my husband." She pointed to our friend, Roger, and I said,

"No, that's my boyfriend." Then I reached to touch Jack and added, "This is my husband." The orderly roared with laughter and the uneasy tenseness in the elevator evaporated. His spontaneous laugh was the last sound I remember before having surgery.

My thymus gland was enlarged. The surgeon said that it was the size of a grapefruit in the shape of a football. The gland was abnormally active. Fortunately, there was no tumor in the gland.

INTENSIVE CARE AND THEN HOME

My thymectomy surgery took over three hours. During surgery, I also had a tracheotomy—a surgical incision into the trachea which allowed me to breathe without complication. After recovery time in the surgical Intensive Care Unit, I was moved to a larger ICU which had about sixteen beds. While this area was staffed with a number of nurses, emergencies were frequent and there were times when I felt alone in the room. I was fed through a tube and a number of other tubes were hooked up to me. I could not speak but I was able to communicate by writing messages.

Unless you have been in an ICU, you may not be able to imagine my frustration and uneasy feelings. I felt isolated from the world and had no control over what was happening to me. It was never longer than several minutes before a nurse would check on me but it often seemed like an eternity. In some ways my body was remarkably strong. At the same time I was as fragile as a snowflake. There was also the constant worry of things that could go wrong—especially in the stressful atmosphere of an ICU.

I could sense the love from my husband as well as his anxiety. He wanted so desperately to "fix things" and see me smile again. Being a corporate executive, he was used to analyzing and solving problems. But no amount of flow charts could change the course of my situation. The support of Jack and my family during this critical time made me realize how blessed I was. I was also grateful to my doctors who were progressive and moved quickly to remove my thymus gland. While this type of surgery is a fairly common procedure today, it was a rare operation in 1978.

Death is a stark reality when you are in ICU. The medical personnel are careful to shield patients from the mortalities that occur but they can't do so entirely. I was the only myasthenic in ICU where patients were coming and going on a daily basis. A young man in his twenties had been in an automobile accident, a teenager was injured in a ski accident, and an old gentleman was there after a stroke. Several heart patients were admitted. A number of the heart patients had been shoveling snow. I have a vivid memory of a young woman and a middle-aged man dying the day after their arrival.

One patient in a cubicle across the room from me had been in ICU for three months. He had more tubes hooked up to him than I did, and it was obvious he could not speak and was too weak to ring the bell for a nurse. One day I looked across the room and one of his tubes had slipped out of place. There was no nurse in sight. He was gasping for breath and the look in his eyes was frightful. I kicked the end of my bed to make enough noise to get the nurse's attention and pointed to the man in trouble. Once he was hooked up again, he gave me a big smile of thanks. Perhaps help would have arrived without my assistance. I only know that it gave me a boost to help someone else while I was feeling so weak. God works in mysterious ways.

I stayed in ICU for twelve days and then was moved to a private room. The sunshine outside mirrored the happiness of regaining some semblance of normality. My tracheal tube was removed and I was able to eat food again. At first I had to place my hand over the hole in my throat in order to speak or eat. Eventually, the hole closed and gradually my speech became clearer. I was comfortable with the knowledge that it was temporary.

What a happy moment when my doctors told me I could go home before Christmas. My daughter made a large "welcome home" sign, and my sister came from Texas to join the family celebration. Our home was decorated with three Christmas trees and Jack even remembered to order the turkey for Christmas dinner.

To add to the joy of the season, the street we live on was (and is) lavishly decorated for Christmas. (Our street was even featured in a book by Mary Edsey entitled, *The Best Christmas Decorations in Chicagoland.*) The 200 block of Claremont holds the record for the longest running group display. In her book, Ms. Edsey includes a photo of our decorations, describing our tradition in this way:

"A matching Christmas tree trimmed with multicolored bulbs on the lawn of every house has been a holiday tradition

since the end of World War II. In the years that have followed, the custom has spread throughout Elmhurst.

On a block so organized it has a mayor, secretary, and treasurer, it is no wonder that uniform precision can be achieved. The homeowners purchase the trees in bulk and place one in front of each home, ten feet from the walkway. Christmas banners, made by one of the residents, hang from lampposts, and "toy soldiers" designed and painted by the residents are placed at both ends of the one-block long street. A five-pointed star crowns the center of the street, hung from the few remaining elms that formerly shaded the entire avenue.

The holiday tradition begins with a private block party—no matter what the weather. It features homemade cookies, hot cider, horse-drawn sleigh rides, and a visit from Santa. Each family brings a log for a blazing bonfire and all join in singing carols. This enduring custom of Claremont Street has kept alive a neighborhood spirit many in our bustling lives have yearned to rekindle."

To be home for Christmas with my family and to enjoy the "Claremont tradition" was truly a blessing. It was also the beginning of a new phase of my life. My "mystery guest" had finally been unmasked and the preliminary step of surgery was completed. Now I needed to recover and begin treatment in earnest, an important stage in my journey to wellness.

POSTSCRIPT

So many advances have been made in surgical techniques since my thymectomy in 1978. Recently, two of our friends successfully experienced robot-assisted surgery for prostate cancer. DaVinci is a computer-assisted robotic system that expands a surgeon's capability to operate (in certain areas of the body) in a less invasive way during laparoscopic surgery. The DaVinci system allows greater precision and better visualization compared to standard laparoscopic surgery.

Operations with the DaVinci system are performed with no direct mechanical connection between the surgeon and the patient. The surgeon is remote from the patient, working a few feet from the operating table while seated at a computer console with a three-dimensional view of the operating field. Patients with this type of surgery have a much shorter hospital stay, less pain, rapid recovery, and early return to work compared to patients with open surgery procedures.

Robot-assisted thymectomy is now being studied and compared with conventional thymus surgery in MG patients. Dr. Jens C. Ruckert of the University Medicine Berlin-Charite, Berlin, Germany, is leading a group of scientists analyzing data from the two types of surgery. Preliminary data points to improved results with the DaVinci surgery but more testing and comparison of the two approaches need to be evaluated. Perhaps this will be a "giant step" in the treatment of myasthenia gravis.

CHAPTER 3

THE AFTERMATH OF TRAUMA

A major operation followed by twelve days in the stressful atmosphere of the Intensive Care Unit was an emotional experience of such intensity that it left a lasting impression. Watching people around me die will always be imprinted in my heart and mind.

When I returned home, I felt as if my once energetic body was trapped in an exhausted body with two settings: all or nothing. Much of the time it was "nothing" but when I got a burst of energy I gained assurance. This surge of vigor seldom lasted long and sometimes led me into an unreal world where I tried to deny the destructive process of my illness. Yet each experience—the times of complete exhaustion when I felt depleted, the lapses into denial, the moments of invigoration and fresh hope—has been valuable and helped me to grow in new and unexpected ways.

POST SURGERY

I found that recuperating from surgery was a very slow process. Week after week went by with little change in my speech. This outcome was not entirely unexpected. My doctor had told me that improvement usually took months or even years before the potential benefit of the thymectomy would be apparent. This uncertainty and baffling vagueness seemed to be a common thread woven into my "mystery guest."

When I returned home from the hospital, I was greatly relieved. It was so good to be in my own bed. Even the familiar sounds of the clock ticking, the refrigerator recycling, and the coffee percolating were comforting and reassuring. And I was heartened by the loving support of my family and the professional expertise of my doctors. Yet soon after returning home from the hospital, I became fearful and distraught. I focused on myself, anxious for my life to return to normal. Eventually, my thoughts centered on the fact that it's ultimately my responsibility to get well and survive this ordeal.

I began to realize that recouping from a chronic illness is different from dealing with a short-term illness. A broken bone is healed and life gets back to normal. Even many heart attack patients can, with proper care and diet, resume a normal level of activity. In my case, I recovered from surgery within a few weeks but my chronic illness, while not life-threatening, may never go away.

Of course, I am aware that research continues and someday a cure or a new drug may be found. Still there is the struggle of coping with a chronic illness along with the ups and downs, good and bad days, valleys and mountaintops. I wanted to be the healthy and vibrant person that I once was. I needed my independence, energy, and sense of wholeness. I longed for physical activity, and I missed the joy of sharing social gatherings with my friends. Most of all, I yearned for a better quality of life. And I had many questions on how to reach these goals.

First, I asked myself, "What is the worst case scenario?" I am recouping from surgery and have an autoimmune disease. There is no cure. How should I approach recovery? And how do I deal with the curiosity of those around me? When someone asks how I am, should I respond, "I'm fine" or is it a lie? The truth is there is no necessity to directly answer that question.

If I am at my grandchild's baseball game and he whiffs the ball, I might say, "Enjoy the game, you will get that home run someday." The reality may be that my grandchild has no athletic talent but do I need to express that truth at that moment? When a friend or a stranger asks, "How are you?"—does that person really want to know? Often I reply, "Fine." Does it make a difference? If I reply, "I have a terrible headache or I'm in pain," and continue with the gross details, what will be that person's reaction? Perhaps he or she will walk the other way the next time we meet or, if trapped, avoid the question.

Many of my acquaintances were (and are) unaware of my illness. Some of my friends assumed that I was in remission after surgery. Often I smiled to hide how I felt. If I can put a smile on someone's face or try to bring some cheer into another person's life, we both feel better.

I have talked to other myasthenics about their experiences in dealing with MG. One of the comments we frequently hear is, "But you look so well!" I've heard it often. While it is meant to boost morale, it also implies doubt that we are indeed ill. Although a myasthenic may look the picture of health, symptoms of muscle weakness and rapid fatigue are not always visible and can happen without warning. We may feel great one minute and like a wet noodle the next. At the same time, we need to participate as "normally" as we can in life's events.

Empathy is needed rather than pity. Empathy conveys an understanding of feelings, whereas pity suggests distress in the perception of suffering. A touch of the hand or a sincere hug can bring more comfort than any words or advice. While being a good listener is an important part of offering support, we also value friends who allow us to vent our anger, frustration or bitterness and then offer honest feedback.

In my own situation, I found it important to talk to my husband and close friends and admit, "I feel tired, my neck and right leg are weak and I need to rest them." Sometimes eating and speaking were difficult, and I realized how poorly I was functioning. Making decisions seemed monumental. Once I was active and energetic, an avid golfer and tennis player. Now I was debilitated both physically and mentally because of the muscle weakness and fatigue of MG. With all of these thoughts swirling through my head, I realized the need to express my feelings and keep a record of my progress for myself and my doctors.

KEEPING A JOURNAL

A journal seemed to be an ideal way to track my progress and determine which medications were most beneficial. Most importantly, a journal would give me a way to express my emotional state during setbacks, and help me to find common patterns in my journey.

I used an 8x10 monthly calendar and wrote in it every day for almost two years. I measured good and bad days on a scale from 0 (the worst) to 10 (the best). When I felt I was making no progress in getting better, I would browse through my journal and realize my life was gradually improving. I began to notice common patterns in attaining certain objectives.

By setting goals I was directing attention away from myself and striving for definite aims. I knew I had limitations and could not raise the bar too high, positioning myself for disappointment. My goals were focused on slowly regaining my health so I could deal with my chronic illness with as little medication as possible and improve the quality of my life.

I determined that goals must be specific, measurable, attainable, and tractable. My journal would help me to make a list of goals and follow my progress. Some of the aims of my journal are:

1) To be specific: A general purpose was not sufficient. I needed to state exactly what I wanted to achieve.
2) To measure: I would set a time limit of two weeks, months or more.
3) To attain: I needed to have a plan with realistic goals to avoid disappointment.
4) To track my progress: I would follow my course by noting advances.

Recording even small improvements in a journal is a source of encouragement. It takes more than *wanting* to regain health. I needed to be specific about a quality life and give myself time to reach each goal, one step at a time. I discovered that small steps can bring noticeable results. Gradually the feeling of having some control over my life gave me hope that the quality of life I longed for was attainable.

Keeping a journal has helped me in so many positive ways. I found it a godsend for my emotional and physical well-being. Most of all, it enabled me to focus on some of my immediate needs following my diagnosis and surgery.

What was necessary for my recovery? Spiritually, I needed to place my trust in a loving God and seek His support. Emotionally, my family and friends were a big boost to my positive attitude. Their encouragement and understanding during those "lean" years were essential ingredients in my continual pursuit of wellness.

Physically, I needed to assist my neurologist in finding the medication or combination of drugs which would give me the most benefit. This process would often require adjusting dosages. Finding an exercise program and a nutritional plan tailored to my needs were also basic concerns.

Then too, it was essential to learn more about myasthenia gravis. Knowledge would empower me to deal with MG. It was also important for me to know the signs of a myasthenic crisis and how to be prepared for it, as well as finding ways to manage the stress of a chronic illness. There was lots of work to be done in the aftermath of my thymectomy and these were the areas I tried to tackle.

NURTURING FRIENDSHIPS

Ultimately, the special relationships I have with my friends played a big part in my recovery. No matter how smart or independent you are, everyone needs somebody to confide in, to listen, and to understand without exploiting and judging.

My husband is my best friend. We have a bond unlike any other and I cherish our friendship. But I also need women friends to sustain and support me.

I recently read a book, *I Know Just What You Mean* by Ellen Goodman and Patricia O'Brien. It's a great story drawing on their long-term friendship. They tell how friends anchor you without judging. And they discuss the differences between men and women friends. Here are some of their findings (paraphrased) about true friends:

"True Friends"

"Are supportive, nurturing, understanding, caring;
Are not jealous of others' accomplishments;
Listen, are loyal, you can tell them anything;
Enjoy the pleasure of your company;
Usually have something in common;
Need not agree with me, yet are there for me;
Share good times and bad times."

And the difference between women friends and men friends:

"Women friends want to be together;
Men friends want to do together.
Women base friendships on shared feelings;
Men base friendships on shared activities.
Women share troubles and relationships;
Men do things together.
Women accept more than men accept;
Women want shared values;
Men want shared interest.
Women discuss problems, happy times, and laugh;
Men discuss sports, combat, army, tell stories, and laugh."

Do you recognize some of these patterns in your own male and female friendships? I mention this delightful book because my friends were so important to me, especially during my postoperative care. My women friends were incredible. They knew speaking was difficult for me. When I could have visitors, they came in groups of six or eight, usually bringing their own refreshments. They happily talked about what was happening in the news and in their lives. I had the benefit of listening without the stress of trying to talk. Holding a conversation was difficult when one person came to visit.

My friends also brought meals when I returned home. We could have ordered out or my husband and teenage children could cook, but it was a big relief when a friend brought our dinner.

I continue to appreciate my women friends. Ours is a mutual bond of being caring, supportive, nurturing, and understanding. We are not jealous of the accomplishments of others. We share the good times and the bad times, and we continue to truly listen to one another.

There is no mystery in my friendships. My close friends know me well. They tell me I am reliable, consistent, and loyal. They admire my discipline, persistence, and patience. I hope I can always live up to their high opinion and be a true friend to each of them.

Nurturing my friendships has become more of a vital concern since I was diagnosed with MG. I must confess that living with my chronic illness has also changed me and my priorities in other ways. Instead of carefully planning my week, I have learned to "wing it" (my daughter's favorite expression) most of the time. Yes, I am guilty of multitasking on occasion but I try to focus on living each moment with everything I do and every person I see.

TREATING MY MYSTERY GUEST

My medication also played a large part in my treatment and recuperation. From the beginning my neurologist explained a number of treatments, some having extensive side effects over a long period of time. He told me that the combination of certain drugs and surgery has been quite successful in the treatment of myasthenia gravis, allowing many patients to lead fairly normal lives.

The first line of treatment for MG is usually a category of drugs called anticholinesterase agents that strengthen neuromuscular transmission. While these drugs don't repair the basic deficiency of acetylcholine receptors, they do prolong the effects of the acetylcholine messenger that signals the muscles to contract. Treatment with anticholinesterase agents is beneficial but not sufficient to allow most patients to return to full activity. A thymectomy, the removal of the thymus gland is often the next step. Thymectomy is performed to increase the chances of remission in the future, not to improve muscle weakness in the short term.

In my own case, my neurologist recommended an immediate thymectomy. Although I noticed a marked improvement, it took time to regain my strength. Drug therapy gave me a much needed boost. Initially, my doctor prescribed Mestinon™.

According to the Myasthenia Gravis Foundation of America, Mestinon (pyridostigmine) is the most commonly used anticholinesterase for the symptomatic treatment of myasthenia gravis. Mestinon prevents the breakdown of the neurotransmitter acetylcholine, allowing more acetylcholine to be available and accumulate. Acetylcholine (ACh) is the chemical which transmits nerve impulses to muscles. With more acetylcholine, there is more control and strength of voluntary functions such as opening and closing eyes, eye movements, swallowing, breathing, and moving arms and legs.

Mestinon did help to alleviate my MG for awhile but it soon proved ineffective. As my body became conditioned to the drug, increasingly larger doses were necessary to relieve my muscle weakness and overall fatigue. In addition, I was plagued by a number of side effects which included nausea, abdominal cramping, and insomnia. A change was needed.

After a few months of discussion concerning the pros and cons, my neurologist prescribed prednisone, a synthetic hormone commonly referred to as a "steroid." Prednisone is chemically similar to cortisone which our body manufactures and is often given to patients with severe, generalized myasthenia gravis as well as many other illnesses.

When you agree to take prednisone, you also consent to run the risk of some complications. Basically, the steroid drug prednisone is classified as a powerful immunosuppressant, anti-inflammatory drug used to gain control of various diseases. Its potential side effects can be quite serious and destructive. Because prednisone inhibits the immune system, there is the danger of contracting and not being able to fight off infections. Other side effects are a false sense of well-being, nervousness, insomnia, weight gain, indigestion, restlessness, fluid retention, bruised skin, excessive hair growth, muscle cramps, skin rashes, and a tendency for the face to become somewhat round (moon-faced) in appearance.

The laundry list of side effects continues. When the drug is taken over a long period of time, prednisone can aggravate and/or trigger cataracts and glaucoma. It can activate a peptic ulcer leading either to perforation or bleeding from the ulcer. Prednisone can hasten osteoporosis or thinning of the bones. While some bone thinning is part of the natural aging process, particularly in older women, prednisone exacerbates the process without regard to age. Post-menopausal women taking prednisone have an added risk of bone fractures and vertebrae collapse.

Because prednisone acts as a mental stimulant, patients have to guard against euphoric feelings and the overly optimistic thought that they can conquer the world. When I was finally well enough to attempt doubles' tennis thanks to prednisone therapy, I boasted to my opponents, "Today is a prednisone day, so you had better watch out!" I definitely felt stronger on the days I took prednisone.

(Another cautionary note about prednisone: anyone who stops taking prednisone must do so gradually. Suddenly discontinuing the drug can be disastrous. The body needs time to adjust to a decrease in the medication.)

Given all the information on the risks of prednisone, I was monitored carefully when I began to take the drug. I began with large doses and gradually reduced the dosage to 20 mg every other day. Over a period of time, I experienced many of the side effects mentioned above. Yet I *was* improving and the benefits of this medication far outweighed the risks. Still, it made me realize how important it is to understand the benefits and risks of any medication prescribed by my doctor.

Many years later, my doctor reduced my dosage of prednisone to 10 mg every other day. This was considered a maintenance dose with fewer potential side effects. My physician did prescribe other medications from time to time without success. The process of trying new medications took almost a year for each change to be evaluated. For fifteen years I was on a maintenance dose of prednisone. This dosage of medication combined with my diet and exercise program enabled me to feel good much of the time. No doubt this sense of well-being encouraged me to continue taking prednisone. However, the long-term use of steroid medication is never completely safe or risk-free. Today, MG treatment is far superior than management programs offered decades ago. If a patient is diagnosed with myasthenia gravis and has access to quality treatment, the disease can usually be managed and the patient can lead a relatively normal life.

In the aftermath of my thymectomy, it became apparent that I must learn everything I could about my disease. Knowledge is not only vital in combating an illness, it's also empowering. I soon discovered that there were other learning hurdles to climb.

RESPONDING TO AN MG CRISIS

One of the first things a myasthenic is told is to avoid emotional upsets since MG is worsened by stress. My doctor and my own investigations led me to learn about the most feared complication for myasthenics — an "MG crisis." This term is applied to the patient who develops an abrupt worsening of muscle weakness, especially a weakness in breathing. It is feared because an MG crisis can kill. A myasthenic may be perfectly healthy for months at a time, and then be nagged by symptoms or floored by the devastating onslaught of an MG crisis.

There are two different kinds of life-threatening respiratory crises, depending on the cause of the distress. If the crisis is brought on by an exacerbation of the disease itself, the patient may require artificial respiration and immediate transfer to a hospital. This sudden spike in disease severity may occur during menstruation, viral or bacterial infection, excessive physical activity, pregnancy, emotional stress, surgery or trauma.

Crisis may also occur because of an overdose of anti-MG drugs. In a cholinergic crisis, an overdosage of anticholinesterase agents may cause the patient to develop increased weakness. You may remember that patients taking anticholinesterase medicines become conditioned to the drug and often have to increase the dose to maintain its effectiveness. Too much of the drug, however, can lead to a cholinergic crisis. When breathing is seriously impaired, treatment must begin immediately to avoid a potentially fatal situation.

Whatever the cause, an MG crisis is always an emergency and should be dealt with quickly. Deterioration can progress at an alarming rate and can lead rapidly to death from respiratory muscle weakness. Efforts must be made immediately to assist the patient's breathing. The airway should be cleared of mucus and saliva and kept open, with the chin up, head tilted back, and jaw forward. Artificial respiration (CPR) should be undertaken, if necessary, until the patient is transferred to a hospital.

Newly-diagnosed myasthenics or myasthenics hospitalized for medication adjustment or other medical/surgical conditions are potential candidates for crisis. While the crisis is temporary, the goal of treatment is to keep the patient alive. When the aggravating circumstances subside, the patient usually returns to the pre-crisis state. In order to avoid a myasthenic crisis, the early treatment of infections, judicious management during surgery, attention to emotional factors, and patient education about medications are essential.

My doctor emphasized the main points I needed to keep in mind if I were to experience a "crisis."

1) I should be thoroughly familiar with the warning signs of an MG crisis.
2) My family and friends should know how to handle the crisis while waiting for transport to a medical facility.
3) I should realize that a crisis lasts for varying lengths of time and stress will worsen the condition.

4) I should keep my identity card (furnished by the MG Foundation, giving my medical history), my MedicAlert tags, and bracelets easily accessible—making it easier for a new doctor to help.
5) I should stay calm! (Read on.)

MANAGING STRESS

Did you know that over two-thirds of office visits to physicians are for stress-related illnesses? Stress is a major contributing factor (directly or indirectly) to coronary artery diseases, cancer, respiratory disorders, accidental injuries, cirrhosis of the liver, and suicide—the six leading causes of death in the United States. Obviously, stress can have both physical and emotional effects on your health and well-being. However, by identifying and taking control of some of the stress in your life, you can lessen the negative effects.

The first thing I realized is that stress is *my response* to an event, not the event itself. Events that are stressful for some may not be difficult for me. For example, a morning commute may be stressful for my friend because he or she spends that time worrying. The same commute may be relaxing for me because I can read a great book or listen to my favorite music.

When I find myself dealing with a stressful event, I notice a number of physical changes in my body. My heart beats faster, blood pressure rises, breathing quickens, facial muscles tighten, and my body perspires. I know I need to relax by thinking of the positives in my life to ward off the harmful effects of ongoing stress.

For most of us, it's easy to be aware of stress in our daily life, but it's a challenge to know how to change it. Given the seriousness of an MG crisis and the damaging effect stress can have on my illness, I knew that I needed to make significant changes in my lifestyle. I had to find a way to reduce the sources of stress and learn to cope with the stresses that could not be eliminated. In my journal, I itemized some of these changes:

1) Exercise and eating well are necessary components. (I discuss each of these topics in later chapters.)
2) Prioritize—plan and pace my activities.
3) Sufficient sleep and rest are essential.
4) Patience! Improvement in my health will take time. Serenity during the process will help reduce anxiety and stress.

Stress that triggers harmful symptoms seems to be a common component in all autoimmune diseases. The experts generally define stress as *any activity that requires coping*. I remember reading about Hans Selye, a Canadian physician who popularized the idea of "stress without distress." While in medical school in the 1920s, he studied the causes of stress. He described stress as the "nonspecific response of the body to any demands upon it." (Selye died in Montreal at the age of 75 in October, 1982.) His research led him to identify a General Adaptation Syndrome which has three stages:

(1) Alarm;
(2) Resistance;
(3) Exhaustion.

The *alarm phase* is similar to a whistle signaling a message: *Wake up and be alert; there's danger ahead.* Our bodies recognize the alarm and a whole series of physical reactions begin. Hormones released by endocrine glands trigger our hearts to beat faster and our lungs to breathe more rapidly. We perspire more, our blood sugar levels increase, pupils dilate, and the digestive system slows down. All of these changes prepare us for our "fight or flight" response to the stress. Even though there may be no physical threat to life, the body reacts as if there were.

As the stress subsides, we then proceed to the *resistance stage* in which our bodies work to repair and stabilize the damage. But what happens if the stress continues with no relief in sight? Then we're in a state of alertness and our bodies cannot undertake the necessary repair work. If this condition persists, we then enter *exhaustion,* the third stage of stress. And, if exhaustion lasts long enough, all sorts of maladies may develop. Headaches, irregular heartbeat rhythm, digestive difficulties, and even mental illness are some of the unhappy consequences of long-term exhaustion. The body's energy resources are spent. Stress weakens the immune system leaving our bodies more vulnerable to illness.

So if this is how stress works, what can we do about it? First of all, we must recognize that no one is completely free of stress nor is all stress harmful. We may find ourselves in the middle of worrying arguments and the demands of a hectic schedule. All of these things produce responses of anxiety, uncertainty, frustration or anger which can result in headaches, rapid heartbeat, and soaring blood pressure. But when stress becomes excessive and puts the body in a constant prolonged state of being tense, good health can be destroyed. Studies even show that our daily requirement of nutrients increases remarkably under stress and energy stores can be quickly depleted.

How do we respond to stress? We all know some people who do better than others in coping with stress. Handling stress is inevitably an individual thing because of the variations of personality, environment, attitudes, and experiences. What is unpleasant for one person may be enjoyable for someone else. No matter what the stressful situation is, we can maintain good health by converting stress into a beneficial motivator. Rather than allowing stress to become a detrimental force, we can make stress work *for* us instead of *against* us.

To my mind, the key concept in dealing with stress is *control*, organizing a sense of power over your life. You cannot control things that happen to you, but you can learn to manage how you respond to them. Coping with the reality of myasthenia gravis in my life, I have learned that I *can* control how I respond to my "mystery guest," a visitor which shows no signs of departing from me.

Not everyone reading this book is a myasthenic. But all of us have to deal with stress issues. It is important to find a purpose in life that suits your own personal stress level. Most of us fit into one or two categories. There are the "racehorses" that thrive on pressure and are only happy with life in the fast lane. Then there are the "turtles" that require a more paced, tranquil environment—a situation that would bore or frustrate the racehorse types. There are also those who try to be racehorses by charging through life as if it were a competition for first place. This can be emotionally exhausting.

In order to control your emotions, it's good to recognize situations as either life-threatening or non-life-threatening. Discover how to respond rather than react. Most confrontations in life are with imaginary predators. We often cause many of our problems. It is better to learn to adapt and live with difficulties than to react in a state of alarm and resistance. I believe negative emotions are a waste of energy resources. They keep you from positive thoughts that release brain endorphins which make you feel better. Emotionally upset individuals often withdraw all of their energy reserves ahead of schedule and can literally run out of life too soon.

RECIPE FOR COPING

I realized that getting enough sleep and rest were vital for me to cope with the stress of dealing with a chronic illness. Lack of sleep can lessen the ability to deal with any stressful situation. When I am tired, I definitely become more irritable. And when I am deprived of sleep for long periods, I often found that even the simplest of tasks was nerve-racking. I would get jumpy and short-tempered. I felt mentally distracted, made mistakes, and found it difficult to concentrate. Over time, I've learned that myasthenics require more sleep than most people, usually ten or more hours a night.

I have discovered I can be productive with the discipline of a schedule, calendar, notepad, and a basic routine. But I also must allow time to "smell the roses" and sometimes sit long enough to watch the roses grow. At the same time, I must prepare myself to handle unexpected crises or disappointments by knowing as much as possible about MG and stress. Listening to my body and learning to observe the warning signals has fortified me for the long haul. And I have realized the importance of avoiding self-medication. The only drugs I take are prescribed by my physician. Drugs or alcohol may seem to mask symptoms, but they don't help me accept or adjust to my situation; many are addictive and self-destructive.

Worrying about health problems is a waste of the energy needed to cope with living a quality life. When life situations are difficult, I think of the problems solved and challenges overcome. I recall the setbacks that eventually moved me into a different direction. Each phase of my life has been a learning experience.

Most importantly, I take time each day to connect with God and give thanks for my blessings. The reward of this link to the Almighty is a sense of inner peace, a peace that *surpasses all understanding* and enriches my life. This serenity has enabled me to be patient with the inevitable setbacks and steady in my emotions.

When you find the peace of God's grace, everything changes. You begin to see beauty instead of ugliness, love instead of hate, and health instead of sickness. The world becomes as fresh as the cleansed air after a rainfall. And, if you look intently, you may even see a rainbow.

I bring peace with me by the attitude I have daily. Many times I have found myself getting caught up in frantic attempts to change things in my life to find a certain sense of harmony. But I allow nothing to disturb my peace because ultimately nothing matters except my connection to God. I have learned to be content whatever the circumstances. Of course I prefer "no pain" to "pain" and I love it when everything goes right in my day. While I may be dealing with at least a dozen health issues, I am grateful I can do most things in moderation.

As I look back, I am aware of a Divine Spirit within me that has graced me with the wisdom to resolve problems and strength to survive tough times. Some people call that intuition or credit their "angel's" guidance. Whatever it's labeled, it's there...I am not alone just as *you* are not alone.

The stress of having MG has encouraged me to seek medical attention and make changes in my lifestyle. While these approaches have helped me in the healing process, I believe the true source of my healing is the renewing power of God within me. With God's spirit at the very core of my being, my body can mend from the inside out.

Lastly, I have found that laughter is the best tension breaker. I try not to take myself too seriously and work hard to avoid the plague of self-pity. It helps to remember that the measure of a life is not the length of it but how it is spent.

I hope you will begin your journey by first accepting where you are in the aftermath of trauma in your life. And then I encourage you to develop and practice your belief in a Divine Spirit greater than yourself. Have faith in your spiritual journey. Follow it if you can, change it if you must, but in the words of Winston Churchill, "Never, never, never give up."

POSTSCRIPT

Current issues with worldwide terrorism make my journey with a mystery guest seem almost insignificant. Trauma, uncertainty, and emotional upheavals play a role in all of our lives. How we respond to crises and stress demands all of our resources.

Many people seek healing and wellness yet they focus on sickness and suffering until they establish that image. Their thoughts and words produce a vivid blueprint, and they live within the bounds and limitations of that design. I have worked at directing my thinking and conversation to generate a positive blueprint in my life. You can do the same.

Today, my neurologist tells me I am "off the charts" in dealing with MG as well as cranial nerve damage from my 2001 brain surgery. Sometimes he has medical residents from Loyola Medical School observing. He smiles at me as he describes the many ways I have worked to have a better quality of life. What a boost to my self-esteem to have such support and encouragement!

CHAPTER 4

THE SUBPLOT OF ACCEPTANCE

One of the treasured moments in the life of a newborn is the baby's first smile and laugh. True, it may be just the passing of a gas bubble and have nothing to do with anything we would consider funny. Oh, but the parents and grandparents are in awe. *Where's the camera? We have to capture this!*

No one knows exactly why we laugh or why people find different things funny. What we do know is that a good laugh makes us feel happy and when we feel good, the world is a rosier place.

Is it possible to use laughter as a means of restoring health? Of course, a serious illness is never funny. Yet I found that having a little fun explaining an incident in my illness was effective in talking to my grandchildren. After my last surgery (removal of a meningioma) I had lost a lot of weight and was still recouping when we visited my son in Wisconsin. I was looking down at my six-year-old grandson when he said, "Grandma, can I ask you a question?" I replied, "You can ask me anything, Michael." Then he asked, "Why does your neck look so yucky?"

My son was shocked and looked at me with sympathy. I tried not to laugh. I said, "Michael, part of what you see is because Grandma is getting older. However, because I lost a great deal of weight and muscle tone after my operation, the skin on my neck is loose. That is why I am working hard to eat well and exercise to rebuild my muscles."

He smiled and ran off to play with his brother. This was a chance to share with my grandchildren how physical changes affect those around us. Today I can still chuckle thinking about this incident and the innocence of a child. Now that he is older and I am healthier, he thinks I should join him, his brother and father in earning a black belt in karate. For now, I will settle for yoga.

What does the current research say about the benefits of laughter and humor? Paul McGhee, PhD has written several books about laughter and humor: *Health, Healing and Amuse System* (1999), *Understanding and Promoting the Development of Children's Humor* (2002) and *Small, Medium at Large: How to Develop a Powerful Verbal Sense of Humor* (2003). He believes that introducing humor at an early age helps children accept and cope with life.

When I was first diagnosed with MG, I recall reading Norman Cousins' *Anatomy of an Illness*. Cousins described his struggles with a painful and life-threatening degenerative illness and his successful self-treatment with vitamin C and episodes of funny film clips such as Candid Camera, The Marx Brothers, and

Laurel & Hardy. His creative approach was an inspiration and motivated me to press forward with *accepting MG.*

ACCEPTING WITH GRACE

Accepting the fact that I have myasthenia gravis was not easy. Webster's dictionary describes the word **acceptance:**

"To take with good grace, to change life style, to acknowledge as received, to compensate for what is lacking."

Even though I had no choice but to accept the reality of my illness, the question was "How?" I made up my mind to rise above the fear of the unknown by learning as much as I could about myasthenia gravis. Perhaps I could even teach others by my example. For me, there is no "right" or "wrong" way to accept what I cannot change. I needed to find my path to deal with and survive my illness and, in the process, grow as a person, mother, and wife.

It helped me to realize that although MG is a serious, autoimmune neuromuscular disease with potentially life-threatening symptoms, most patients can be medically controlled. Myasthenics no longer need to be referred to as "victims." While there is no cure for MG, research and education have made great strides since I was diagnosed almost 30 years ago. And with continued efforts, a cure may eventually come.

Once I accepted my doctor's treatment I began implementing a plan. As my medicine helped to control my disease and my body regained strength, my type "A" personality geared into action. I was ready to go forward with my goals to live a quality and productive life. The key for me was to persevere and not settle for second best.

I know doctors are busy and I also recognize they are concerned about me. But as time went on, I questioned my doctors about treatment options. You may also find it helpful to actively participate in your medical care. If you are not able to do this, find a family member or friend to be your advocate.

As a person with this disease, I am unique. No one else can live my life, understand what I feel or experience the world as I do. I believe it is important to maintain a happy, positive outlook and not become "victimized" by my thoughts. I have had to redefine my expectations and, until a cure is found, learn to live with MG. I have chosen to reconcile myself to what I cannot change. As I began accepting that my disease was part of my life, I could envision a future. I made a list of what would help me accept my mystery guest:

1) Define what I was going to accept.
2) Realize that my lifestyle will change.
3) Stop focusing on the things I cannot change.
4) Recognize the importance of my attitude in determining the course of my disease.
5) Educate myself, my family, and my loved ones about my disease.

6) Learn to cope with fluctuating physical limitations.
7) Allow myself to accept my life and find peace.

To my mind, acceptance offers a sense of freedom and inner peace. I refuse to be a "victim" and become ineffective in managing my disease and, therefore, my life. By developing healthy habits, I've been able to participate in my well-being. I believe this is true of anyone with a prolonged illness. Healthy habits are learned in the same way as unhealthy ones, by practice. My options may be limited but doors of opportunity are opened.

I was thankful that I had a name for my health problem. I no longer had to speculate on what will or can happen. Life is a journey of experiences that can lead to the "mountaintop." I needed to find a way to make this journey a blessing. I could not dwell on what I lack but rather be thankful for what I have.

Fundamental changes needed to be made not only in lifestyle, but also in expectations. Life is what you make it. I found that it often makes little difference what is happening in your life; it's how you *accept* what's happening that counts. Facing an illness is a turning point in anyone's life. The way it is handled can make our lives richer and fuller. I recall the words of Reinhold Niebuhr:

"God grant me the serenity to accept the things I cannot change, the courage to change the things I can, and the wisdom to know the difference."

65

FAMILY MEMORIES

Where does one find the strength to accept adversity and persevere? My early family experiences set the stage for my adult outlook on life. I grew up in a small east Texas town, Pittsburg, Texas. I am distantly related to the Pitts family. In 1855 when much of our great country was first being settled, a wagon train carrying the families of H.A. Pitts and his brother W.H. (Major) Pitts traveled southwest to the area known today as Pittsburg, Texas—named in honor of this pioneering family. My father, John Gaftner Matthews, was a direct descendant of the Pitts family.

My mother's family, the Lakes, moved to Camp County in the Pittsburg area in the early 1900s. Bessie Lake Hood married my father in 1927. Their union was blessed with five children while raising cattle, cotton, potatoes, feed for the cattle, and the most delicious and biggest watermelons in the county.

Shortly after the stock market crash that triggered the Great Depression and Europe was on the road to World War II, I was a baby without a care in the world. By the time I became aware of the reality of the Depression, I was a teenager.

On our 300-acre farm in Pittsburg, there were horses to ride, cows to milk, and pigs to feed. For many years my father was a farmer. I didn't know we were poor and in many ways we weren't. Actually, we were fortunate to grow our food and process our meat which protected us from much of the hurt that others endured during the Depression. I had many good friends and neighbors, the security of wonderful parents, and an extended family of grandparents, aunts, uncles, and cousins. (For awhile, my grandmother Lake and my grandfather Matthews lived with us.) In many ways, it was an idyllic childhood.

My parents deserve credit for nurturing a love of learning and encouraging our natural curiosity. My mother was a teacher and our home was filled with books. My brothers and sisters and I were never bored; we had a thirst for knowledge and a passion for life.

We were taught to perform any task and do the best we could, both mentally and physically. I can still recall some of the admonitions that became a part of my life: *Go that extra mile and whatever happens in life, deal with it. With strength and discipline, you can do anything. Add positive thinking to that and watch things get better.*

LEARNING ACCEPTANCE AT HOME

I was eight years old when my father became ill. The doctor found a goiter (an enlargement of the thyroid gland) and surgery was recommended. The nearest hospital for this surgery was in Dallas, Texas, about 125 miles away. Leaving my mother and four children behind (my oldest sister was away at college) my father took a bus to Dallas. An aunt and uncle who lived in Dallas met his bus and took him to the hospital where the goiter was removed. When my dad left the hospital, he stayed with my aunt and uncle who took care of him until he was well enough to return home.

Dad's tumor was not malignant and there were no complications. But this was the beginning of my knowledge of autoimmune-related diseases in my family.

In my generation, children were to be clean and well-behaved, seen but not heard. I was not told much about my father's illness. He recovered well and life went on as usual.

(As an adult, my sister Evelyn had symptoms which indicated problems with her thyroid. She had an underactive thyroid gland, a condition called hypothyroidism in which the gland does not produce sufficient thyroid hormone. Evelyn's hypothyroidism was regulated with medication. The year before my illness was identified as myasthenia gravis, I was also diagnosed with hypothyroidism.)

In 1967, I got a phone call from my brother who told me my father was very ill, dying of lung and stomach cancer. My husband and I had moved to the Midwest the year before. Our children were two and four years of age. I took the children with me to Texas to see my father and comfort my mother. My husband arrived a few days before my father died.

Recently, I found a copy of my grandfather's obituary which read in part:

"One of the county's most honorable citizens and a successful farmer died Saturday morning. No man in the county was more highly regarded for integrity and religious fervor than Mr. Matthews. He was influential in every cause involving moral principles and good government."

My first thought was *this could describe my father*. My Dad's courage, discipline, vision and endurance were identical to his father. Dad would often say,

"Without character your accomplishments are nothing at all," and "If it's worth doing, it's worth doing right."

Dad was a proud man of the finest character, someone to admire. He always gave one hundred percent of himself to his family and friends, and often to total strangers. Dad died of cancer years before I was diagnosed with MG, but his strength has stayed with me. His legacy of always striving to do one's best has helped me to live a much more fulfilling life. Losing my Dad at such a young age was difficult and it was not easy to accept.

My father-in-law, who we affectionately called "The Duke," had a similar philosophy. He said, "If you work hard, you can accomplish anything in life." He was a good-hearted man who gave away much of his fortune during the Depression to help needy people. Later he worked hard to regain his former affluence by his persistence and determination. ...The Duke is a constant reminder of how I should live my life.

By their example, my parents and other family members have taught me to accept what I could not change. I hope and pray I can do the same for my children and grandchildren.

THE ROLE OF FAITH IN ACCEPTANCE

Some people think you have to choose between trusting science and placing your faith in God. I don't believe this is true. Science and faith are two separate ways of dealing with issues and can work together. In the case of a chronic illness, science provides medical answers to restoring health; faith helps us deal with the uncertainties. To my way of thinking, science and faith complement one another.

Martin Luther King, Jr. once described faith in a powerful and practical way—"Faith is taking the first step even when you don't see the whole staircase."

My faith is strong. I know that with God's support I can meet the challenges of each day and nothing is impossible. Where does this gift of faith spring from? My family has played a big part in my faith. My parents had a code of ethics which they taught to their children. They allowed us to learn from our mistakes by giving us the freedom to choose. They showed us love as well as declaring it.

My mother was a woman who lived her faith. By the strength of her own commitment, she taught me that whatever the situation I must accept it and then make the best of it.

I remember her last trip to visit us in Illinois. American Flight #360 from Dallas to Chicago arrived on time. I waited eagerly for the door to open and the passengers to appear. When I saw her, she looked as if she had aged since our last get-together. Her white hair was freshly groomed but the character lines in her face seemed to give her an anxious look. Still she walked proudly with a straight back and the assurance of knowing who she was and where she was going. I looked at her kind brown eyes, and when she saw me the worried look was erased. Her smile brought back many memories.

Certainly mother was a strict disciplinarian but she always found creative ways to make work fun. Her laughter could fill a room much to the delight of everyone around her. She taught school and later raised five children, and was a frequent volunteer for classroom mother, field trips, and teaching bible school in the summer.

At age 86, mother continued to be active. She was secretary of her church which had 800 parishioners, secretary of her Seniors' Sunday School class, and active in her XYZ (extra years of zest) Club. When there was a parish dinner—which was often in her small town—she was usually the one in charge. She would explain, "It's something I can do for my church."

She loved people and people loved her in return. Even in her later years when she was plagued with many ailments, she volunteered at a nursing home and documented the library's town history by interviewing the "old timers" in the community. Friends and family marveled at all she accomplished. For my dear mother, it was a stimulating and faith-filled way of generous giving.

Her spirit and zest for life was contagious. She made you feel like you could do anything. She was a remarkable person and I was proud to call her my mother. Although she died several years ago at age 98, her memory still inspires and strengthens my resolve to place myself in the hands of a loving God.

My serious illness gave me an opportunity to use and put my faith into practice. While many people subscribe to a belief system, I wonder how many also draw on their faith and make a concerted effort to develop their religious life.

The late Dr. Robert Laaser, past minister of St. Peter's Church in our Chicago suburb, often said, *"Using your faith is the only thing that makes it grow."*

I believe the Lord sometimes warns us to slow down and take better care of ourselves, but we don't always listen. I didn't heed the message until I was very ill and barely able to speak or eat. I thank God that He gave me another chance. Often when an illness occurs, people say, "Why me?" Or "It is the will of God." I can honestly say that I didn't question or think, "Why me?" And I positively do not believe that any illness is the will of God.

Leslie D. Weatherhead deals with this subject in an eloquent way in his stimulating book, *The Will of God* (1944; revised in 1972, 1999). He divides the subject into three parts:

1) *The intentional will of God—God's ideal plan for man;*
2) *The circumstantial will of God—God's plan within certain circumstances;*
3) *The ultimate will of God—God's final realization of His purposes.*

Let me explain how I interpret these ideas on God's will. God's *intentional will* is demonstrated in the first two chapters of the Book of Genesis which tells of perfect goodness. (The Book of Revelations ends with a similar idea.) God *intends* for us to be healthy, strong, successful, and live our lives with abundance. Poverty, loneliness, hatred, pain, sickness, and hunger are contrary to God's intentional will.

We know that bad things do happen. People suffer from diseases, loved ones die, human tragedies occur. How does God's will fit into these difficulties? Certainly they are not part of God's intentional will.

Pastor Weatherhead offers the idea that there is a will of God within tragic circumstances. When bad things happen, God's *circumstantial will* helps us to adapt to the evil conditions on earth. In my own case, God is working in all the circumstances of my life and giving me the graces I need if I turn to Him with confidence and faith. There is an awesome mystery in trying to understand how God allows misery and suffering. But I cling to my hope and trust in God to see me through the dark hours.

Lastly, there is God's *ultimate will* that we should live our lives so that someday we will be reunited with God in heaven. As a child I memorized the answer to the question, "Why did God make us?" The response was, "To know, love, and serve God in this world and be happy with God in the next." These simple words are profound and require faith.

Although a positive attitude and courage are admirable qualities, it is our belief system which fulfills us spiritually. God promises to make use of the circumstances in our lives to serve His ultimate will. And evil things in life do not reflect God's intentional or ultimate will.

Denis Waitley, author of *Seeds of Greatness* (1986), talks about faith and perseverance. He writes,

"Faith and perseverance are similar, but different in that perseverance is the test of faith."

Perseverance is also "hanging in there" when the odds stack up against you. A belief system filled with determination allows you to bend and grow with the winds of change without breaking your spirit. It helps you to look at the bright side of the darkest situation.

So I draw strength from the examples of beloved family memories and the goodness of a loving God. Life does give us "mountains to climb." The mountains don't go away, but faith, acceptance, and perseverance help us to climb them.

OPTIMISM AND EMOTIONS

Political journalist/editor Norman Cousins was the author of *Anatomy of an Illness as Perceived by the Patient* (1979, revised edition 2005), *The Healing Heart* (1984), and *Human Options* (1991). He writes,

"Optimism is an incurable condition in the person with faith. Optimists believe that most disease, distress, dysfunction, and disturbance can be cured. Optimists also are prevention oriented. Their thoughts and activities are focused on wellness, health, and success."

The writings of Norman Cousins reinforce the idea that a positive attitude is essential in order to achieve and maintain a good, healthy life. And I wholeheartedly agree.

My mother-in-law, Hazel, epitomized a positive outlook. She wrote me bright, cheerful notes daily while I was in the hospital. Like my own mother, Hazel's optimism stemmed from her deep faith in God. She truly believed that humor is the best medicine one could take. Hazel's advice to *keep a sense of humor in all that you do* saw me through many an ordeal with (I hope) a smile on my face.

Hazel's encouragement to research and write about my experiences with MG motivated me when the going was rough. After reading an early draft of my manuscript, Hazel (in typical fashion) suggested I eliminate the boring technical part and replace it with humor. A good idea perhaps...for my next book. I do believe cultivating a sense of humor sustains you when your health is at stake. Before Hazel died, she made me promise that I would finish my book. Little did I know how long it would take to accomplish that goal.

I also learned from my mother that my mental attitude can make the difference between enjoying life and being overwhelmed by it. Worrying over things you can't control can keep you from dealing effectively with the things you can manage. So I try to think positively and get the necessary rest to face the challenges of each day.

You have to be aware that as your life changes your requirements may change also. Allow time for adequate rest, not just sleep. Take time to read for pleasure, write to a friend or just put your feet up and let your mind wander. Don't procrastinate until tomorrow. Start now...but do continue reading!

Emotions such as love, pleasure, joy, fear, anger, and grief play a major role in maintaining a positive mental attitude. Whatever is happening in my life, emotions shape my perception and response to illness.

In my own case, I had not anticipated my emotions being so powerful. Suddenly, I was unable to do many of the things I had done before and became dependent on others. Needless to say, for a naturally "in control" and organized person I was not a "happy camper." It was difficult to distinguish whether my emotions were becoming a problem or my reaction to my illness needed help.

After the initial relief of being diagnosed, I experienced a certain amount of anger. Anger is a tricky emotion. Screaming in desperation was pointless. Life may not seem fair but all around me were examples of people with more problems than I had. On the one hand, I felt justified in being upset that a doctor did not diagnose my illness sooner. However, if I allowed anger to take hold, it could prevent me from progressing with the physicians treating me. Besides I didn't think I was truly angry with my situation and I never succumbed to the "Why me?" attitude.

On some level I believed things just happen in life. At times I was disappointed and I was definitely frustrated with my increasing dependence on my family. It was an effort to not vent my frustrations on others. Even though I knew things could be worse, I was not in control of my life. Over time, I recognized that what I was feeling could be *displaced anger.*

I knew my anger was not directed at someone or something. But I was exasperated at my inability to be physically active and do things for my family, and I missed my friends and social life. At the same time, I knew it was unhealthy to stay angry, keep my anger inside or give vent to outbursts that would cause stress and aggravation.

While I prided myself on dealing with fears and keeping myself motivated, I also needed to acknowledge my frustration. It became clear that my best interests were served by putting anger aside and forging ahead with thoughts of getting well.

Dealing with a chronic illness drains a person physically, mentally, and emotionally. No doubt my acceptance of my illness passed through various stages of anger and denial. If denial is an abnormal refusal to accept the truth, I am sure that there were times when I fit into this pattern.

Learning to reconcile the emotional overtones of a chronic illness is a lifetime journey. I have learned to respect my feelings and realize that how I act is more significant than how I feel at any given moment. The key to coping and acknowledging my emotions (good or bad) requires a willingness to educate myself about my illness and emotions, while practicing habits which give me the best quality of life. This is a slow, steady process that offers a level of independence and reduces my anxiety and fear. When my emotions are intact, I feel better and my body is in a healing mode. And, if my body is in a healing mode my medication is more effective.

I decided I wanted to enjoy the moment without the uncertainty of tomorrow. Each day is different; symptoms fluctuate from hour to hour or at least from day to day. My limited energy and stamina makes planning ahead difficult. Sometimes I have a bad day. I can't smile and say everything feels fine. On my poor days, I have learned to allow for *downtime,* giving myself permission to feel bad, sad or even helpless.

To counter my gloomy days, I try to take a mental vacation. The best part about this technique is that I am only limited by my imagination. For example, I like the snow (a good trait for a Chicago suburbanite) so I visualize walking in the snow. I listen for the crunch as I walk, sense the warmth of my coat and boots, and feel the invigorating tingle of the cold air on my face. Using my imagination, I can go anywhere. The key for me is not to let downtime extend into days or weeks, and slip into self-pity.

Using this visualization strategy has worked for me. Some people prefer talking to a counselor or a friend. Whichever technique or combination of methods is used, don't let disappointments, unpleasant emotions, anxieties or regrets deflate you for long periods of time. Whatever helps you to overcome the downtimes, do it and then move on.

THE POSITIVE PATH TO ACCEPTANCE

After my diagnosis, I was advised to avoid emotional upsets which worsen the symptoms of MG. But what about positive emotions? If negative emotions produce harmful chemical changes in the body, was it possible that positive emotions, such as love, hope, and joy, could have a therapeutic effect on body chemistry? I recalled the story Norman Cousins described in his book, *Anatomy of an Illness.* Cousins' awareness of the deleterious effects of negative emotions led him to dwell on positive emotions. His good health was eventually restored.

I gradually realized that I must be in command of my emotions at all times. I had to replace anxiety with confidence. I had to slow down, whether I was doing the laundry, walking to the library or even talking. This relaxed pace was not easy because it was contrary to my usual fast and sometimes frenetic mode of operation.

When I was first diagnosed in 1978, I wondered if there was a scientific basis for the belief that *positive* emotions have beneficial effects on the body's chemistry and functions. Norman Cousins certainly believed in such a mind/body connection. Today, numerous scientific studies have confirmed the link between emotions and physical well-being.

While the biochemical effects of negative emotions (e.g., panic, anxiety, fear, suppressed rage, exasperation, and depression) are associated with adverse physical aftermaths, the opposite is true of positive emotions. It is now well documented that positive emotions can actually *block* the destructive nature of panic or fear which can intensify illness. No doubt the positive emotions serve a specific purpose in protecting the human body in sickness and in health.

You may be aware of the role of blocking agents in medications. Some of these substances are designed to protect the body against biochemical activity such as an erratic heart rhythm. Other drugs are designed to intercept severe pain in joints and muscles, while others modify the chemistry of the brain in mental illness.

I think the beneficial and reversing action of blocking negative emotions *by turning on positive feelings* is similar to these therapeutic medications. Of course, striving to maintain positive emotions may not work in all circumstances. But the potential benefit is sufficient to boost one's confidence in the necessity for making special efforts.

With my faith and a positive mental attitude, I have confidence that I can turn negative emotions into positive ones. I have confronted the mystery guest, myasthenia gravis. I have "accepted the things I cannot change" and will continue to work toward "changing the things I can" as God gives me "the wisdom to know the difference."

POSTSCRIPT

My family has been a great support in helping me to "accept what I cannot change." No one said life would be easy. There are times when I feel overwhelmed and there seems to be too much to accept. However, if I can muster the courage to face the challenges of each day, I feel that I am on the right path.

I am reminded of a favorite saying of my father's: *"Acceptance takes character and character is what you do when no one is looking."*

His example of courage, discipline, vision, and endurance has guided me in my own quest. All things seem possible with a positive attitude and an understanding of God's will in the overall scheme of life.

CHAPTER 5

SEARCHING FOR CLUES

Isn't it great to curl up with an absorbing mystery book? *It was a dark and stormy night...* The criminal deed is usually only a small part of the drama. Much of the story describes the investigator's discovery of key pieces of information. Eventually a solution is reached and the culprit is apprehended or otherwise eliminated.

Something similar happened to me in my quest for wellness. Once my body recovered from surgery and my mind and spirit accepted the reality of MG, I was free to explore. My aim was not only to learn more about myasthenia but also to search for ways to fully participate in my journey to well-being.

There is an old and wise saying, "Charity begins at home." In my own voyage of discovery, I began with myself and my family and our physical battles with autoimmune diseases. Then I branched out into research concerning the effects of nutrition and the mind/body relationship in health. Along the way I met and learned about some inspirational people who changed my life for the better. But let me set the stage of what was happening when I began searching for clues.

****

It was 1979 and Jimmy Carter was President of the United States. As the decade of the 1970s neared its close, many things were changing. According to New York Yankee fans, it was the year of "The Bronx Bombers." It was also the year Pope Paul VI died in the fifteenth year of his papacy. The cost of a college education had tripled since the mid-1960s. In two years, our children would be entering college. The world's first test tube baby was born in England. And another form of reproduction called cloning or using genes from any cell to create an offspring dominated the headlines.

Another important event that year was the inability of the United States Congress to agree on legislation to regulate the controversial DNA or gene-splitting research. Most scientists agreed that the new gene-splitting technique constituted a powerful research tool with extraordinary potential. Gene research today has come a long way and is still evolving. (You'll find more discussion of gene research in Chapter 12.)

Myasthenia gravis is an autoimmune disease. It is *not* a genetically transmitted disease. However, it has long been noticed that there is an unusually high incidence of certain genetically determined blood and tissue types among MG patients. It seems probable that specific genes or gene combinations place certain people at a greater risk of an autoimmune disease. So my first priority was to understand more about the immune system and autoimmunity.

THE BODY'S DEFENSE SYSTEM

The immune system is a group of organs and cells located throughout the body, including the thymus, spleen, bone marrow, lymph glands, and white blood cells. Normally, the immune system detects and attacks foreign substances such as viruses and bacteria. In MG, the immune system's antibodies mistakenly attack the acetylcholine receptor sites on voluntary muscles that receive messages from nerves.

While scientists know that antibodies are the problem in MG, it's still not clear how the immune system is stimulated to make these antibodies. Most researchers think that the development of MG probably involves several factors.

Although MG is not an inherited disease, it is possible that having certain immune system genes increases a person's likelihood of developing MG when other triggers are also present. Some possible triggers that stimulate the autoimmune response are viruses or infections of any kind, extreme stress, drugs, and environmental pollutants.

Many researchers believe that a further risk for MG arises due to certain abnormalities in the thymus gland. The function of the thymus is to screen and destroy cells that can react against the body's own tissues. If this screening fails to filter out cells that produce antibodies to the muscle receptor sites, the stage is set for an autoimmune reaction. Because of the active role the thymus plays in the immune system, medical scientists have long suspected that thymic abnormalities are an important factor in developing MG.

It is believed that genetic susceptibility and thymic abnormalities probably combine with other still unidentified agents to trigger MG. The Myasthenia Gravis Foundation of America has funded pioneering studies into the causes of MG, and continues to lead in the quest for further understanding of the complex details.

The term "autoimmune disease" refers to a varied group of more than 80 serious chronic illnesses that involve almost every human organ system. It includes diseases of the nervous, gastrointestinal, and endocrine systems as well as skin and other connective tissues, eyes, blood, and blood vessels. In all of these diseases, the underlying problem is similar—the body's immune system becomes misdirected, attacking the very body it was designed to protect. Some of these diseases and the organs or systems attacked are:

- Addisons's disease—the adrenal cortex.
- Aplastic anemia—the bone marrow.
- Autoimmune hepatitis—the liver.
- Celiac disease—chronic inflammation of intestinal villi with gluten intolerance.
- Crohn's disease—the small intestine.

- Diabetes mellitus (Type I) — the islet cells of the pancreas.
- Graves' disease — the thyroid gland (hyperthyroidism).
- Guillain-Barré syndrome (GBS) — the peripheral nervous system.
- Hashimoto's disease — the thyroid gland (hypothyroidism).
- Multiple sclerosis — the central nervous system (brain and spinal cord).
- Reye's syndrome — the brain, liver, and other organs.
- Rheumatoid arthritis — the joints and sometimes other organs.
- Systemic lupus erythematosus (SLE) — any organ, principally skin, kidneys, joints, and heart.

Although many autoimmune diseases are rare, there are numerous diseases which fit into the autoimmune category. As a distinct group of disorders, they affect millions of people.

While myasthenia is an autoimmune disorder, AIDS is classified as an *acquired* immune disorder. According to Dr. Richard Tindall, a member of the National Medical Advisory Board, MG is the opposite of AIDS. People with MG produce too many destructive antibodies, while AIDS patients have a weakened immune system which produces too few antibodies. The vast amount of AIDS research is leading to a better understanding of the immune system. These studies as well as research on MG are mutually beneficial.

FAMILY TIES

Since I was diagnosed with an underactive thyroid gland before MG was identified, I began my research by learning about the genetic component associated with hypothyroidism. The thyroid is an endocrine gland in the neck with two lobes, one on each side of the windpipe. It is well supplied with blood vessels and takes in iodine from the blood. When the iodine combines with other chemicals in the gland, thyroxine, an important hormone is formed. Thyroid hormones are necessary for body growth and mental development.

My father had a goiter, an enlarged thyroid gland which may be accompanied by either hypo- or hyper-thyroidism. Goiters develop when there is not enough iodine supplied to the body for the thyroid to function properly.

When the thyroid gland is inactive as in hypothyroidism, everything slows down and there is a loss of vigor of mind and body. Body weight is gained due to slower metabolism. Doctors treat a hypothyroid condition by giving the patient thyroid hormone supplements, usually in the form of a pill.

In a hyperthyroid condition, the thyroid gland produces too much thyroxine. This may be due to overactive gland cells or an enlarged gland with an increased number of cells producing thyroxine. In either case, a superabundance of thyroid hormone results in hyperthyroidism. The signs of hyperthyroidism are extreme nervousness, increased heartbeat rate and metabolism. In treating hyperthyroidism, surgeons often remove part of the overactive gland. Radioactive iodine is sometimes used to decrease thyroid gland activity.

When a thyroid disorder is left untreated, elevated cholesterol levels, heart disease, depression, osteoporosis, and infertility can occur. Because of these possibilities, early detection is critical. Doctors can now detect thyroid disease in early stages when symptoms are subtle or not even present. The tool for this diagnosis is a simple blood test known as the sensitive thyroid-stimulating hormone (TSH) test.

It's interesting that my father, my sister Evelyn and I have all had autoimmune thyroid problems. Was there a definite genetic pattern of vulnerability in our family history? Other family members have also experienced bouts with autoimmune diseases.

JOHN AND REYE'S SYNDROME

At age 12, my son John was diagnosed with Reye's syndrome, an autoimmune disease. Anyone who lived in the Chicago area in 1973 will remember the epidemic of Reye's syndrome. Television and newspapers were filled with stories of dozens of cases at Children's Memorial Hospital. A picture of one 13-year-old boy, Barry Olander, was on the front page of the Chicago

Tribune. Barry was in a coma for three weeks. Fortunately, he survived.

That same year, my son John was home from school with a cold that progressed into a flu-like illness with a high fever. It was the week before Easter. As a family, we have memories of John often serving as an acolyte, especially for the Maundy Thursday, Good Friday, and Easter Sunday services. This Holy Week and Easter season, John was home with a high fever.

Earlier in the week I took John to the doctor's office. His symptoms did not seem serious. He was given aspirin and told to get rest. My husband Jack left for the Good Friday service. I stayed home with John and our ten-year old daughter, Susan.

Things were going well and John seemed to be feeling better when suddenly he became violently ill. As his fever spiked, he became lethargic and began to vomit so intensely that he was producing green bile.

At seven o'clock in the evening, I called our pediatrician Dr. Russell Snow. He told us later that he knew I was not a mother easily alarmed so he said that he would meet us at our local hospital. I notified Jack at church and he rushed home to take us to the hospital. Our daughter Susan went to stay with a friend and neighbor. By now it was obvious that our son was very sick.

When we arrived at the hospital, Dr. Snow ordered a blood ammonia test. The level was so high that Dr. Snow called a colleague at Children's Memorial Hospital. (We believe that this immediate action saved our son's life.) Within minutes I was in an ambulance with John en route to the Children's Memorial Hospital in Chicago. During the twenty-five minute trip, the paramedics were attentive to John's needs.

I remember looking at him as only a mother can. He had always had a beautiful tan color to his skin. Now his skin was so white there were freckles across his nose. He was awake and I'm sure he saw the concern on my face when he said, "It's OK, Mom." At that moment I believed him because I knew the Lord was with us and would see us through this crisis.

A father's emotions are equally as strong as a mother's when a child is seriously ill. Jack was grateful that his friend Dr. Russ Snow was driving him to the hospital, and took solace in the good doctor's calmness and assurance that he would confer with the medical team when they arrived.

A team of doctors worked on John for four hours, giving him a blood transfusion and relieving the pressure on his brain. In the wee hours of the morning, the doctors entered the waiting room with smiles on their faces. Yes, John had Reye's syndrome but fortunately he was given care during the crucial first stage of the disease. The physicians reassured us that John would survive and make a complete recovery. We never saw this medical team again. After they spent time with us they were off to intervene in the next crisis situation. A new group of doctors and nurses were assigned to John, and he was moved to a room with another Reye's syndrome patient.

As grateful as Jack and I were with the good news, we still had some lingering apprehension. Then we walked into John's room and his roommate hopped out of bed, walked over to Jack, extended his hand and said, "Hi, I'm Barry Olander." Jack shook his hand, thinking he looked very familiar.

Barry continued, "I'm the boy who has been in the news. I was in a coma for three weeks and now I'm fine. If your son is in this hospital, he'll be alright." What a relief to hear that encouragement. We knew all would be well. After all, a 13-year-old boy and a skilled team of doctors agreed on the outcome.

Finally, we were able to relax somewhat and feel comfortable with our good news. John was out of danger and the medical team insisted we go home and get some sleep. Reluctantly we drove home. Our friend, Dr. Snow, had thoughtfully left our car in the hospital parking lot.

When we arrived home we took a hot shower and climbed into bed. It was warm under the covers but we were still tense and unable to sleep. As the morning dawned, we called our daughter to assure her that her brother was doing well. It was an upsetting and frightening time, but Susan understood that we needed to spend more time at the hospital with John.

Four days later we brought John home. The hospital nutritionist gave us a strict six-week nutrition plan, basically a low protein diet. Limiting a growing, pre-teenager to consume small amounts of meat and carbohydrates was not an easy task. Somehow it didn't matter because John was feeling well and we knew he would be even better. Today he is healthy and happy, and leading a productive life with his family of five.

The medical dictionary describes Reye's syndrome as: *an acute illness of children...characterized by an encephalitic-like state with fever, vomiting, disturbance of consciousness progressing to convulsions and coma.*

According to the National Reye's Syndrome Foundation, symptoms of the disease usually occur soon after a child has had a viral infection, especially the flu, chicken pox or upper respiratory infection. The most common symptoms are:

- Persistent or recurrent vomiting.
- Listlessness or loss of pep and energy.
- Personality changes such as irritability, combativeness.
- Disorientation and confusion.
- Delirium and convulsions.

A child can often progress into a coma within an hour and a half after he begins to vomit. When vomiting takes place, progressive organ problems may have already occurred. The liver may malfunction and begin to accumulate fat. The brain swells producing pressure which can sever the blood supply and cause brain damage or death.

It is important to note that Reye's syndrome is not contagious. It is diagnosed through blood ammonia and sugar level tests.

Unlike a lot of other childhood diseases, early detection is the key to saving lives. If a child with Reye's syndrome is misdiagnosed or discharged from the emergency room, it could be a devastating and fatal decision. When Reye's syndrome is diagnosed in stage one (vomiting), the chances of recovery are greatly enhanced.

In October, 1982 I wrote to the National Reye's Syndrome Foundation to ask if there could be a genetic link or vulnerability between myasthenia gravis and Reye's syndrome. I received the following reply concerning these autoimmune diseases:

"At the present time our doctors cannot find any link between Myasthenia Gravis and Reye's syndrome. Reye's is caused by a virus first, plus some type of heredity deficiency, such as an enzyme problem or the environment."

You can find more information about the warning signs of Reye's syndrome and local chapters in your area by contacting:

The National Reye's Syndrome Foundation
nrsf@reyessyndrome.org
1 (800)-233-7393

EVELYN AND ADDISON'S DISEASE

Another member of my family also experienced a bout with an autoimmune disease. In 1991, my sister Evelyn Matthews became seriously ill. Except for severe headaches and an underactive thyroid gland, she had been the "picture of health" most of her life. Suddenly she felt extremely weak, was losing weight, had low blood pressure, and was hospitalized on several occasions due to dehydration. Most of that fateful year was spent with doctors and test after test without a definite diagnosis.

Almost a year after her first symptoms, Evelyn was diagnosed with Addison's disease. Treatment with steroid medication helped her regain her strength and gradually get her life back to normal. I saw Evelyn several times that year. She gained weight and was full of energy.

The following year Evelyn's headaches became more frequent. Often she would go to bed requesting no one to call and disturb her sleep. After two or three days of bed rest, she would be restored to "almost normal health." A close friend checked on her several times a day and made sure she had some nourishment. From time to time when Evelyn went out-of-town, she forgot her medicine and her symptoms of weakness returned. When she resumed her medication she was fine.

In 1993 Evelyn's headaches returned again. We didn't realize she was often skipping her medication when she had severe headaches. The headaches made her sick to her stomach and she could not retain the medicine. She became weak again and I made plans to fly to Dallas.

I talked to Evelyn for twenty minutes around 6:00 PM on January 29, 1993. She sounded good...just tired. Her last words to me were, "I love you, take care of yourself." My daughter, who lived in Texas and talked with Evelyn frequently, tried to call her around 8:00 PM. She did not answer the telephone. The next morning, a friend found Evelyn had passed away during the night. An autopsy determined that Evelyn had died of "complications of Addison's disease." This was not an easy thing for me or my family to accept.

Once again, a member of my immediate family had an illness directly related to the immune system.

Addison's disease is an autoimmune disease that is characterized by chronic and insufficient functioning of the outer layer (cortex) of the adrenal gland. This malfunction results in a deficiency of the hormone aldosterone which regulates the salt and water balance in the body. Patients with Addison's disease show abnormally high concentrations of potassium and abnormally low concentrations of sodium in the blood. These changes in electrolyte balance produce low blood pressure and increase water excretion that can lead to severe dehydration.

Other characteristic traits are a bronze-like pigmentation of the skin, progressive anemia, severe prostration, low blood pressure, diarrhea, and digestive disturbances. It can be fatal.

(You may recall that President John F. Kennedy had Addison's disease. This information was not made public until many years after his assassination. With proper medication, he was able to lead a normal life.)

So my search for clues about my own disease began with understanding my family's vulnerability to autoimmune diseases. Armed with this awareness, I was prepared to delve deeper into a thorough investigation of ways in which I could help myself in my journey with MG.

RESEARCH AND FINDING CONNECTIONS

About six months after I returned home from the hospital, I began spending many hours at the library.

The link between nutrition and neuromuscular diseases kept coming to my attention. As I continued my investigation, I read about and met some interesting people. Some have changed my life. I cannot complete my story without mentioning the influence of these individuals. Discovering their experiences has inspired and empowered me to tell you my story.

In the section on optimism in the last chapter, I mentioned the author Norman Cousins, a person who made a great impact on me. In his classic book, *Anatomy of an Illness,* Cousins relates how he overcame an incurable illness by changing his diet, using humor as an antidote, and visualizing wellness. If you haven't read his book, I highly recommend it.

Cousins was hospitalized in the mid-1960s with an extremely rare, crippling disease known as collagen deficiency. When conventional medical treatment failed to improve his condition and he was diagnosed as incurable, Cousins decided to take matters into his own hands.

After much thought and research, and with the help of his wife and a physician friend, Cousins formulated a plan involving changes in his diet combined with lifting his spirits. While the mind/body relationship has been and continues to be thoroughly explored in recent times, Cousins was one of the first laypersons to implement the idea into a concrete application and then share his findings with a worldwide audience.

Being aware of the harmful effects that negative emotions can have on the human body, Cousins reasoned that the reverse must also be true. To boost positive emotions, he borrowed a movie projector and began viewing the Marx Brothers' motion pictures and old Candid Camera reruns. In his book, Cousins recounts that he "made a joyous discovery that ten minutes of genuine belly laughter while viewing these films would give him at least two hours of pain-free sleep." What seemed to be a progressively debilitating and fatal cellular disease was gradually reversed. In time, Cousins almost completely recovered.

Is it possible that laughter, an activity that sets us apart from other forms of life, is not only "good medicine" but also has healing properties? I like to believe that laughing is not only enjoyable but also therapeutic. I know it has helped me face many of my own challenges.

Another individual who influenced my search for healing was Professor Roger MacDougall, a playwright and former professor at the University of California Theatre Department. An interesting article in the health magazine (*Better Nutrition*, Vol. 39, No. 12) detailed Professor MacDougall's dramatic recovery from multiple sclerosis.

In the early 1950s, MacDougall was admitted for tests at the National Hospital for Nervous Diseases in London and was diagnosed with multiple sclerosis. The various treatments prescribed were ineffective and he later said that "one nearly killed him." During this early period of his disease, Roger MacDougall was unable to recognize faces and was virtually blind. He had such difficulty speaking that he was almost unintelligible. Walking was so hard that he was forced to use a wheelchair for almost ten years.

97

Yet twelve years later, MacDougall was driving a car, writing plays, and even bounding up stairs two at a time! MacDougall attributes his remarkable recovery to a gluten-free and refined carbohydrate-free diet enriched by vegetable oils and vitamins. He claims to have achieved a complete recovery from multiple sclerosis by strictly following his diet. He points out, however, that his progress was unbelievably slow. It took four years before he began to make any recovery. (Most of us would not have that much patience.) During this period, he never had setbacks in a disease which usually gets progressively worse. MacDougall was persistent.

According to Dr. David Patterson, Roger MacDougall looked more fit and was more active (twenty years after his diagnosis) than most men his age. His vision was restored; he could run easily across Hampstead Heath (a mile-wide wilderness area in northwest London). Stairs were no problem. The only legacy of all the years spent in a wheelchair was a slight stoop. Clearly he recovered and he ascribed his diet as the key to his remission.

MacDougall's story sparked my interest in pursuing the link between health and nutrition. This led to my discovery of the prolific writer and nutritionist Adelle Davis, author of *Let's Eat Right to Keep Fit* (1970) *and Let's Get Well* (1972, and revised edition 1988). Targeting the millions suffering from ill health, Adelle Davis advised selecting foods that contain the most needed nutrients for repairing and rebuilding a sick body. As I researched my illness, I latched onto the idea of repairing and rebuilding the weak muscles of my body. So I carefully studied Davis's many nutritional programs to aid recuperation and her guidance for renewed health.

Although I had read many books about nutrition and disease, I had not previously found one that mentioned myasthenia gravis. In her classic book, *Let's Get Well*, Adelle Davis not only mentioned MG, she wrote five pages about it.

As an added bonus of reading Ms. Davis's books, I became acquainted with a remarkable woman with MG, Jean Welch Kempton. Her story, *Living with Myasthenia Gravis, A Bright New Tomorrow* (1972) proved to be a turning point in my own life. It was such an inspiration to me that I'd like to share some of her background with you.

As a young woman studying nutrition at Cornell University in the 1930s, Jean-Louise Welch (later Jean Welch Kempton) began to develop the early symptoms of myasthenia gravis. As is often the case, the signs of her disorder came on gradually and were not identified as myasthenia gravis until the disease had progressed and she was in very serious condition.

At one critical point, she was hospitalized at the Neurological Institute of Columbia-Presbyterian Medical Center in New York City. A young neurologist was concerned about her plight and determined to find something to help her. After much study and research, Dr. William Everts decided to try Prostigmin, a powerful muscle stimulant used in surgery. In a matter of minutes, Jean-Louise was able to open her eyes and shed her masked expression. She sat up in about twelve minutes, got out of bed and walked to the nurses' station much to everyone's amazement.

Unfortunately, this incredible change lasted for about an hour and then the symptoms reappeared. But it was a marvelous beginning to finding a treatment for MG.

A year earlier, in 1934, Dr. Mary Walker was working on a difficult case of myasthenia in a British hospital. By chance, Dr. Walker noticed the similarity of symptoms in her patient with people suffering from curare poisoning. (Curare is a poison Amazon warriors used on the tips of arrows.) Dr. Walker proposed treating the myasthenic with physostigmine, the antidote used for curare poisoning. And the patient responded dramatically.

Dr. Walker's breakthrough with physostigmine and Jean Kempton's experience with Prostigmin were the pioneer efforts to actually treat myasthenia gravis. The positive results with Prostigmin led to the development of other drugs such as Mestinon and Mytelase. The permanent removal of symptoms, however, remains elusive.

THE NUTRITIONAL LINK

Meanwhile Jean Kempton researched her disease, focusing on the value of nutrition. She learned that the B vitamins, especially thiamine, were beneficial to those with muscle/nerve diseases. In addition, she discovered the importance of vitamin C and its relationship to healthy muscles. Research studies showed that mice deprived of vitamin C developed muscle weakness of the hind legs.

Because of the findings on vitamins B and C, Ms. Kempton decided to test the value of vitamins in her own situation. After taking a regimen of vitamin C and some of the B vitamins along with a multiple vitamin for a period of time, Jean began to gradually improve. She was even able to decrease the amount of Prostigmin. As her weakness subsided, she returned to many of her activities.

Fortunately, Jean Welch Kempton spread the word of the nutritional approach in dealing with MG. She wrote a book, *Living with Myasthenia Gravis*, was profiled in Adelle Davis's writings, and spoke at symposiums for myasthenics and the medical community.

In September, 1980, I had the opportunity to attend one of Ms. Kempton's symposiums at the Bowman Gray School of Medicine in Winston-Salem, NC. My traveling companion was the late George Wilson who was also attending the symposium. George was a myasthenic who helped form the first MGFI chapter in Illinois in 1972. It was then called the Greater Chicago Area Chapter of the Myasthenia Gravis Foundation and George was its Chairman.

The conference provided an exchange of information among eight MG patients (from different regions of the United States) who were trying various nutritional therapies. The researchers attempted to trace a common pathway of action and design appropriate research focusing on MG and nutritional therapy.

Actually, Bowman Gray's research on MG began in 1969. The biochemists were intrigued by the experience of a patient with severe myasthenia gravis going into remission after being given vitamin and mineral supplements. Bowman Gray researchers have also done basic cell-level studies of muscles and nerves.

The symposium was a chance for me to share with other myasthenics and learn more about my disease. It was also an opportunity to finally meet Jean Kempton. We had corresponded for several years, and her letters to me were always a great source of encouragement. Her graciousness and fortitude in the face of adversity were inspiring.

At the symposium we heard about the various biochemical approaches being implemented to better understand neuromuscular diseases, including the effects of exposure to chemicals and paint. The twenty-six doctors, researchers, and nurses also queried us—the MG patients—about our case histories. And we, in turn, drew up a list of questions we would like answered as they continued their research. We were reassured that Bowman Gray planned to continue research in the area of nutrition and neuromuscular diseases.

Recently, I contacted the Bowman Gray School of Medicine for an update on their work. Bowman Gray is now the Wake Forest School of Medicine. Research studies on the nutritional approaches to treating neuromuscular diseases are ongoing. And part of the funding for the research comes from the Jean Welch Kempton bequest to the medical school. Jean Kempton died of cancer in 1985.

(You'll find more on the latest MG research in Part II, Chapter 12.)

KEMPTON'S CLASSIC BOOK

In addition to *Living with Myasthenia, A Bright New Tomorrow,* Jean Kempton wrote another book published in 1983 entitled *Immunity, Nutrition and Neuromuscular Function.* This book was written with the advice of physicians and researchers at the Bowman Gray School of Medicine. It's an important book, one of the finest of its kind. Jean Kempton points out:

"The science of nutrition slowly has evolved alongside of medicine. We are now entering an era where the marriage of biochemistry and nutrition is beginning to create a change in clinical medicine. The biochemical aspects of specific nutrients at the cellular level are being examined and will have an impact on the treatment of disease.

Change is a slow process, however. It may take years of research and reorganization of medical school curricula before nutrition will become an important part of patient treatment. We look forward to a day when your doctor will be able to advise you how to keep well by regulating your diet according to your genetic, biological and physiological needs."

Jean's book gives the rationale for the use of certain nutrients as adjunctive therapy in treating neuromuscular diseases. She states, "It is hoped that, like a seed planted, there will be growth and maturing in the minds of doctors about this new approach to clinical practice."

The Kempton book relates how the author, after a prolonged serious battle with myasthenia gravis, became well after adding certain nutrients along with anticholinesterase medications. She led a very active life despite having MG, two radical mastectomies, metastatic bone cancer, and excessive stress. Her book is inspirational and I can only add, "Bravo, bravo."

IMMUNE SYSTEM, STEROIDS, AND VITAMINS

Another insight I gained at the Bowman Gray Symposium was the emphasis on the role of the immune system in MG. The immune system is our "first alert" alarm system, protecting us from foreign elements such as viruses, fungi, bacteria or chemicals. A compromised immune system means that the body's first line of defense is broken. Antibodies that were meant to protect can actually cause harm. In people with myasthenia gravis, the immune system produces antibodies that block the transmission of nerve impulses to muscles, causing muscle weakness.

Learning more about the role of the immune system in MG led me to think about my own vulnerability, even before my symptoms were identified. In retrospect, there were many warning signs that my immune system was not functioning properly. As a child, I had a wasp sting on my eyelid which resulted in severe swelling. As a teenager, my contact with a Portuguese man-of-war caused a severe allergic reaction to the poison.

During my adult years, I've had an underactive thyroid and reacted to frequent throat infections with a nasal tone or loss of voice. I have also suffered from bronchitis and pneumonia. So, in reality, I have had problems with my immune system most of my life. Yet, until my diagnosis with MG, I considered myself a healthy person.

While it was important to focus on the immune system, it was also vital to understand the interaction among the immune system, steroids, and nutrition. I learned that steroids frequently given to myasthenics can *alter* immune functions. Steroids can also damage the pancreas and interfere with the production of digestive enzymes. A lack of these enzymes can cause incomplete digestion and absorption, and result in nutrient deficiencies.

Since I was taking steroids, I decided it would be wise to consult a nutritionist to design a diet and vitamin supplement program. I needed to build up my immune system, counter the side effects of medication, and regain my strength as much as possible. In the next two chapters, I'll tell you how I dealt with this three-edged sword of challenges.

Searching for clues to my illness led me along a positive path of seeking insight and understanding. The knowledge I gained from my research helped me to find a direction to my longing to be as well as I could be—physically, emotionally, and spiritually. The poet Robert Frost expressed it well:

"Two roads diverged in the woods, and I
—I took the one less-traveled by,
And that has made all the difference."
From *The Road Not Taken*

Using the discoveries I have made and applying them to my situation enabled me to lead a quality and productive life. Each of us must find our own road to travel. My hope is that some of the insights I have gained from my research will motivate you to find your own path to wellness.

POSTSCRIPT

Some moments in our lives are pivotal. They are so significant that we think of these times in terms of before and after a certain event. In my own life, meeting Jean Welch Kempton was a turning point in my journey to wellness.

Ms. Kempton's pioneering use of Prostigmin, her dedication to advancing the cause of MG research, and her invaluable writings were an inspiration. She became my mentor, encouraging my fledgling efforts to record my experiences. When she invited me to participate in one of her symposiums on nutrition, she became my friend.

I feel a tremendous sense of gratitude to Jean Welch Kempton. Her support and motivation have empowered me to complete this book.

CHAPTER 6

THE PATH OF NUTRITION

The movie, *Lorenzo's Oil*, tells the story of a young boy diagnosed with a rare and fatal nerve disease adrenoleukodystrophy (ALD). The grief-stricken parents spend countless hours researching the disease and consulting with scientists and other parents of ALD patients. Even though neither parent has a medical background, they miraculously discover a treatment for their son's disease.

Since the disease ALD destroys the fatty protective sheath surrounding nerves, the parents strike upon the idea of supplementing their son's diet with olive oil. As a result, their son Lorenzo and countless other victims are spared the devastating effects of ALD. It's quite an inspiring story and an excellent movie.

My own search for a nutritional guideline was not as clear-cut or dramatic. In fact, it was (and is) a daily and evolving struggle. Fortunately, I tend to find a challenge stimulating, whether it is striving to be a good wife, mother, president of the PTA or a charitable organization or winning at tennis. Now I faced a new hurdle, finding what kind of nutritional plan would help me deal with my mystery guest, myasthenia gravis.

Before I was diagnosed in 1978, my life was organized and predictable. Today, almost thirty years later, I am able to maintain a normal life by taking a small dose of medicine every other day. My activities are somewhat limited and I try to not over schedule my day. Given proper rest, I find that I can do most things in moderation. I have learned to relax and my patience is growing. My priorities and lifestyle have changed with the help and support of my family, friends, and doctors.

I was fortunate that once I found competent physicians, they acted quickly and decisively to stem the progression of MG. Within a year after my thymectomy, I had improved. While my doctors were pleased with my progress, I was far from satisfied.

Although I was grateful for any hopeful signs, the side effects of the medication and my muscle weaknesses—especially in my throat—left me feeling unwell most of the time. Without at least ten or twelve hours of sleep at night and a one or two hour rest during the day or if I talked too much, I would lose my voice and get a hacking cough. Within the first three months of my diagnosis, I gained twenty pounds even though I was eating less. I passed out several times, my circulation was poor, my speech was slurred, and my neck muscles were so weak that holding my head upright was difficult. So...yes, I was better than when I was first diagnosed. But there were also many areas that I hoped could be improved and perhaps even be reversed.

WORKING WITH A NUTRITIONIST

I kept a record of my improvement and refused to believe my situation was hopeless. Slow progress was better than gradual deterioration, and I definitely was not ready to accept any possibility of being bedridden for the rest of my life. It seemed vital to counteract the side effects of my medication. As I described in the previous chapter, my research pointed to the potential benefits of a nutritional approach.

With the help of a nutritionist, my doctor, and a yoga instructor, I embarked upon a program of diet, vitamin supplements, and exercise (Chpt. 7) tailored to my needs. It was a gradual process but eventually I regained better circulation. I lost the weight I had gained even though I was eating more food. (The difference was the kinds of food I was eating.)

My program of nutrition, supplements, and exercise has helped me in many ways. In fact, it is one of the reasons I'm writing this book. I hope by sharing my experience, you—the reader—may also be helped. Whether you are struggling with or caring for someone with a chronic disease or if you are blessed with good health, each of us can be challenged to formulate a wellness program.

As I see it, everyone is responsible for his or her own health. Others can make suggestions, but no one else can eat the foods of value for us or exercise our bodies. For many of us, the health we enjoy is largely of our own making. When we want good health and are willing to work patiently toward that goal, the rewards are often forthcoming.

In my search for a nutritionist, I was fortunate to find an outstanding professional. Betty Wedman, an attractive, petite blond, was a perfect example of "practicing what she preaches." She was a registered dietitian with a nutrition counseling practice in a Chicago suburb.

As a nutritionist, Betty worked with patients referred by physicians and dentists, and counseled them about food habits and individual dietary needs. She gave nutrition lectures throughout the world, explaining the rationale of dietary recommendations for diabetes, hypoglycemia, prenatal and infant care, children's needs and general nutrition.

Betty had previously worked with persons suffering from celiac sprue, multiple sclerosis, diabetes, and many other diseases, but she had not worked with a myasthenic. I became a new challenge for her and she became my inspiration.

At our first meeting, I explained my diagnosis of myasthenia gravis, the thymus surgery as well as my prescribed medications. In addition, I told her that my doctors were fairly satisfied with my progress, but I wanted to feel better. Then I told her about my research and some of the things I thought might be helpful. Betty made it clear that she would not attempt to treat my disease. But she could help me counteract some of the side effects of the medication, rebuild muscle tone, improve circulation, and reduce my weight.

I also consulted with my doctor and kept him abreast of the nutritional guidance I was receiving from Betty. He ordered blood tests (SMAC, CBC, and thyroid) which showed that I was deficient in many areas.

Building on these medical findings, Betty advised me that my medication was robbing me of protein and many vitamins and minerals. The drugs I was taking to relieve my symptoms also changed my metabolism of fats and carbohydrates. Obviously I was in need of a diet geared to correct these gaps in nutrition.

MY NUTRITION PLAN

N.B. A note of caution: I am not a nutritionist or a doctor. The following nutritional regimen was recommended to me after consulting with professionals. This is not a prescription for you, the reader. Everyone has a different body chemistry and lifestyle. What benefits me may not help you and vice-versa. If you decide to pursue alternative therapy, discuss it with medical professionals, do your research, and choose your options carefully.

My nutrition plan consisted of a high protein diet (fish, chicken, lean beef, veal, and eggs) and high fiber (large amounts of fresh fruits and vegetables). Refined carbohydrates could be taken every other day in small amounts. Eventually (almost two years later), a gluten-free diet was added. (Gluten is found in barley, oats, rye, and wheat.) I was also advised to drink at least eight glasses of water a day.

Then there was the question of taking vitamins and supplements. Betty thought that in many instances a good, balanced diet is sufficient, but she believed that vitamins and supplements were necessary in my case. Because I had MG, was taking medications and blood tests revealed deficiencies in many nutrients, I was an ideal candidate. Betty also emphasized that the stress of a chronic illness depleted my vitamin and mineral levels.

Working out a nutritional plan reminded me of an earlier conversation with my neurologist when I had resisted taking any drugs on a long-term basis. After he explained why the medication was necessary to control my disease, I agreed to take the drug. But I was concerned about the side effects even though other drugs counteracted the negatives.

A nutritional approach with diet and supplements rather than drugs seemed like a better way to offset the side effects of my medication.

I began with a good multivitamin with minerals and gradually added other vitamins along with lecithin. (Lecithin is a fatty substance found in all the cells of our bodies. Because of its unique role in my recovery, I'll describe it in detail in the next section.) My regimen of taking vitamins, minerals, and supplements has proven to be very beneficial to my well-being.

If you take vitamin supplements, there are some important things to keep in mind. There are two basic types of vitamins based on their solubility: water-soluble and fat-soluble vitamins. The water-soluble vitamins can usually be taken safely in large amounts because they are eliminated by the body. Only a small portion of water-soluble vitamins are stored in the body. The water-soluble vitamins, vitamin C and the eight B vitamins (biotin, folate, riboflavin, thiamine, B6 and B12) are measured in milligrams.

The fat-soluble vitamins, specifically vitamins A, D, E, and K, can be toxic if taken in large amounts because the body stores them in body fat. These vitamins are measured in international units (IU). Again, it's always best to consult with a nutritionist under the guidance of your doctor.

I learned that each nutrient has its own special functions and relationship to the body, but no nutrient acts independently of other nutrients. Although everyone needs the same nutrients, each person is different in his or her genetic and physiological makeup. For this reason, a person's quantitative nutrient need is unique.

In some instances, there are barriers that prevent the proper assimilation of the foods eaten, resulting in more than the "normal" requirements of certain nutrients. It is wise to check with your doctor and keep in mind that what helps one person may not benefit another.

LECITHIN AND ACh

Lecithin took on a special significance when I discovered that this nutrient is important in the production of acetylcholine (ACh). In 1934, scientists confirmed that acetylcholine is a neuromuscular transmitter. Remember that ACh is the chemical messenger between nerves and muscles. A lack of choline (contained in lecithin) can cause a marked underproduction of acetylcholine resulting in muscle weakness, damage to muscle fiber, and extensive scarring of muscle tissue. Many nutritionists believe that lecithin supplements may help to increase the production of acetylcholine.

My nutritionist explained that lecithin acts as an emulsifier in the blood stream. By keeping cholesterol and fats solvent in the blood, lecithin prevents these substances from attaching to the walls of arteries. Commercially, lecithin is added to food products such as margarine. Lecithin becomes an "anti-splattering" agent when margarine is used to fry foods. In candy bars, lecithin prevents the chocolate and cocoa butter from separating.

Lecithin also plays an important role in maintaining a healthy nervous system. In our bodies, lecithin is found in the myelin sheath, a fatty protective covering for the nerves. In addition, lecithin is a key building block of cell membranes. While all of these roles of lecithin are interesting, the one that caught my attention was the possibility of increasing acetylcholine levels by taking supplemental lecithin.

For about a year, I took lecithin without observing any particular change. Then I went on a trip and forgot to take any lecithin with me. I quickly noticed a big difference. The muscles in my neck became much weaker and I had respiratory problems.

A few days after I resumed taking lecithin, the weakness went away. This same experience happened to me on several different occasions when I neglected to take lecithin. Of course, the recommended dosage of lecithin, as well as all vitamins and other supplements, is highly individual and should be discussed with a nutritionist or a physician. Lecithin is available in foods such as egg yolk, soybeans, and corn, and as a supplement in capsule, liquid or granular forms.

BENEFITS OF A HEALTHY DIET

As I was learning about nutrients, I worked hard to stay on my diet. The results were rewarding. Within three months, I felt better. My circulation improved and some of the side effects of my medication were gone. After nine months, the benefits were even more noticeable. I lost weight even though I was consuming more food, and my circulation further improved. The numbness in my right leg disappeared and my muscle tone began firming.

I continued working with Betty for five years and kept in touch with her for several more years. She is another fine woman who helped me make many positive changes in my life. When I follow her diet program, I feel wonderful. I must admit that I'm only human and I cheat occasionally. But I'm always sorry when I do for I have learned that proper nutrition improves the way I look, feel, and act.

Although the ideal way to attain proper nutrition would be through diet alone, there is another problem related to the aging process. As we get older, especially if a person is unwell or taking medications, maintaining weight is important. Balancing nutritional needs while reducing caloric intake isn't impossible but it's difficult. The quality of the food must count more than the quantity. It's simple to suggest reducing starch, sugar, and fats, and switching to protein, fruits, vegetables, and whole grains. We all want good health. We want our minds to be alert and our emotions and feelings to work for us and not against us. The reality is that it's hard to change firmly established eating patterns.

Unfortunately, we cannot quickly step out of our old eating habits into prefabricated new ones the way we can change clothing. Reforming such habits has to be gradual. The obvious challenge I had to face was to eliminate my bad eating habits and substitute good ones.

While working with my nutritionist, I learned a great deal about myself and how nutrition can help me. It was also a revelation to learn that malnutrition can be a factor in degenerative diseases. Until then I had associated the word *malnutrition* with poverty-stricken people in Third World countries or the ghettoes of underdeveloped countries. Now I discovered that the average American eats virtually no breakfast, has a light lunch, and often gorges on a heavy evening meal almost devoid of nutrients. Malnutrition is not necessarily being underfed, but it does involve making poor choices of food and eating habits.

The effects of malnutrition on our bodies are many. Poor nutrition damages the body as it breaks down muscle tissue and taxes every organ in the body. It can even be fatal in extreme cases. Overall, its damaging effects often creep up slowly. In some instances, it's hidden behind a steady diet of "junk food." Proper nutrition is produced when all the essential nutrients, i.e., carbohydrates, fats, protein, vitamins, minerals, and water, are supplied and used in adequate balance to maintain optimal health and well-being.

The sole purpose of nutrition is to build and maintain health. Ironically, many people become interested in building health via good nutrition only after an illness strikes. Medical dictionaries list hundreds of diseases with unknown causes. Many of them are also considered "incurable." Until research can find the solutions for such diseases, I believe that adequate nutrition is imperative in helping to rebuild health.

Dr. Sheldon Saul Hendler, author of *The Complete Guide to Anti-Aging Nutrients*, says,

"As a nation we are overfed and often seriously undernourished. Dietary nutritional imbalances and deficiencies contribute significantly to the premature deaths of millions of Americans annually. Moreover, our changing world has placed new demands on our bodies, exposing them to environmental stresses and insults that deplete our tissues of protective micronutrients as never before."

He further adds,

"Although consumption of the so-called "well-balanced diet" is still thought by many to supply all the vitamins we need in quantities sufficient for the maintenance of good health, there are certain situations that place people at an increased risk for vitamin-insufficiency states. Among those

at risk are people on low-calorie diets, alcoholics, pregnant women, the elderly in general, surgical patients, users of certain medications, and strict vegetarians. Vitamin supplementation is prudent in those and other cases."

DIGESTION AND BLOOD TYPE

What a revelation it was to discover that taking my medication radically changed the metabolism of carbohydrates and fats! For some time, I had thought of medicines as substances treating my condition. But the fact that the drugs I needed to take *altered* food and nutrient metabolism meant they also interfered with digestion and absorption.

According to nutritionists, merely eating the right foods does not necessarily ensure adequate nutrition. Food must be digested and then absorbed by our bodies before we can benefit. Recovery from any disease is hindered until both digestion and absorption are efficient. Digestive factors include how the food is chewed, the secretion of digestive enzymes, and the condition of the digestive tract lining through which the nutrients pass into cells.

The functioning of the digestive system is often a product of one's personality and constitution. For example, one person secretes too much stomach acid and develops ulcers, while another secretes too little acid and cannot digest food. Still other people may have difficulty absorbing needed nutrients. For these reasons, individuals eating the same diet may end up with dramatically different intakes of protein, carbohydrates or other nutrients.

In my case, I have GERD (gastroesophageal reflux disease) which causes heartburn and acid reflux. Actually, GERD is a term for several digestive conditions with different degrees of severity. It can often be managed with a combination of simple lifestyle changes and medication. In rare cases, surgery may be necessary. Because of GERD, I feel better when I eat a diet high in protein, fresh fruits and vegetables, and low in refined carbohydrates and acidic fruits.

I have also discovered that my blood type is a factor in determining the kinds of foods that make me feel better. Dr. Peter J. D'Adamo writes about this connection in his book, *Eat Right 4 Your Type* (1996).

According to Dr. D'Adamo, different blood types denote different blood chemistries. He believes that the secret to healthy, vigorous, disease-free living might be as simple as gearing your food intake to your blood type. His ongoing studies are built on thirty years of research conducted by his physician father. Dr. D'Adamo's book was one more breakthrough in my nutritional research which I heartily recommend to you.

In recent years, Dr. D'Adamo's book has helped me to continue a personal plan that began with my nutritionist. It was interesting to me and worth noting that his strategy is in many ways similar to the plan my nutritionist prepared for me two decades ago. I was inspired by Dr. D'Adamo's research on blood types. The connection between blood type and diet is a new idea for most people, but it often resolves some dietary questions.

According to Dr. D'Adamo: *"Your blood type is the key to your body's entire immune system, and as such is the essential defining factor in your health profile. Your blood type reflects your internal chemistry. It is the key that unlocks the mysteries of disease, longevity, fitness, and emotional strength. It determines your susceptibility to illness, the foods you should eat, and ways to avoid the most troubling health problems."*

My blood type is O-negative and my husband's blood type is A-positive. Our diets vary according to the suggestions given in D'Adamo's diet plan. Both of us believe this dietary change has proven successful.

Do you know what your blood type is? Most people don't. When a woman is pregnant or if someone is preparing for surgery, a blood test is taken to identify the blood type. If a blood transfusion is needed, the correct blood type information is vital.

After major surgery in 2001, I suffered unusual fatigue. I lost a lot of weight, muscle tone, and bone density. To recover my weight and strength I began eating indiscriminately. Three weeks later, I regained my lost weight but I was still extremely fatigued. I needed more sleep, my neck was weak, my stomach was bloated, and my whole body was sluggish. My neurologist often told me not to blame everything on MG. Yet I assumed my myasthenia gravis had been exacerbated by the stressful surgery.

Eventually the doctor prescribed tests which showed that my MG was stable and my thyroid was normal. In light of my downward spiral, my internist and neurologist suggested decreasing my medication for MG. With my weak symptoms I was reluctant, but since cutting back on my medication was a goal of mine I began the slow and long process of reducing prednisone—one milligram at a time over a period of four to six weeks with each reduction. After a few months I felt the same—no better, no worse. Then I thought of a change in my eating habits that occurred after surgery. In order to gain weight and restore my strength, I began to eat indiscriminately without regard to my nutritional plan. As a result, I slipped into some poor eating habits.

I believe eating the wrong food is like putting the incorrect grade of octane in my car. Instead of fueling the engine, it clogs the works. My nutritionist had moved out-of-state and I was not ready to find a new one. Then I remembered reading in *Eat Right 4 Your Type* about several of Dr. D'Adamo's patients with similar health problems. When they tried their "blood type diet plan" their energy returned and their overall health improved. I was not totally convinced that this would help me but I decided to try it. After all I had been through, it seemed simple enough to get back on a diet plan or—as I like to say—"create new and better eating habits."

Deciding to make a change was easy; the difficult part was the execution. I soon realized I had fallen into habits of snacking on sweets, eating chips and French fries with a sandwich, and increasing my intake of wheat and dairy products. Changing these poor nutritional patterns would be another hurdle to overcome but I was determined to do it.

Of course, I was aware that many factors in my life contributed to my disease and my present sluggish body. It would be overly simplistic to suggest that my blood type is the sole determining factor in any improvement. Not all scientists subscribe to the link between blood type and eating habits. In fact, the world of nutrition is full of controversy. Major disagreements exist among experts and many questions are unanswered, particularly on the necessity for supplementary vitamins and minerals.

Within a few weeks of adhering to my blood type diet and returning to my nutrition plan, there were obvious improvements in my energy level, stomach bloating, and circulation. Most of my symptoms were resolved or at least stabilized. Six months later I felt great.

My life is now full of energy and I firmly believe my diet is the key to this change. My doctors agree. The beneficial effects in my life are overwhelming.

Success is contagious. My family has also been helped by forming better eating habits. My husband is a borderline diabetic. Since initiating our more healthful diet, he has been able to control his glucose level by the food he consumes. Jack rarely eats sweets and has decreased the intake of carbohydrates.

There is an interesting sidelight to our improved eating habits. Jack and others in our family have noted that by establishing good eating choices, their systems become more "fine-tuned." When the "wrong" food is ingested, their healthier bodies often rebel in the form of indigestion, hypertension, sluggishness, and weight gain. It seems as if our bodies become more sensitive to poor food choices.

Spreading the word about good eating habits has been a special joy for me. Good friends of ours, Anne and Jim, who own several stores which primarily sell fresh fruits and vegetables, asked me to write a monthly nutritional report on the benefits of specific produce. Each month a free flyer featuring the fruit or vegetable of the month is distributed to their customers. Of course, I realize that I cannot change everyone but surely I can encourage others by my life and example.

FINAL THOUGHTS ON DIET

Improving health through nutrition is a slow process. Certainly it is not as dramatic as stopping an infection with antibiotics or finding an olive oil treatment for ALD. You must exercise patience. The importance of nutrition took years to be recognized, and it takes a long time to correct poor nutritional habits. Of course, the best advice is to begin with good nutrition as a maintenance approach to wellness and the prevention of ill health.

Diseases caused by nutrient deficiency can be improved if it can be determined which nutrients are missing and then adding the needed nutrient to the diet. While this idea is an established one, we need to expand this view to people suffering from any disease process. Patients need to consult a physician and then tailor their dietary needs, preferably with a registered dietitian.

During the last several decades, nutrition has developed into a highly technical science. I believe that nutrition may someday be recognized as the most important factor in treating convalescence from almost every disease. Fortunately, doctors today have a greater appreciation of nutrition than they did in previous years. In this age of specialization, it is encouraging to know that physicians are working with dieticians in rebuilding health.

It is not easy to select the best foods—those which contain the most needed nutrients. In the United States, we produce more food than we need. To keep excess food from spoiling, it is highly processed and refined. Such processing destroys many nutrients and adds chemical preservatives. This is one of the reasons vitamin and mineral supplements are sometimes beneficial.

I do not claim to have marvelous cures. Nor would I ask you, the reader, to follow my nutritional program. What I do suggest is that you work with your doctor and a registered nutritionist to formulate a nutritional plan tailored to your needs.

Unfortunately, illness prevents the body from absorbing necessary and beneficial nutrients. I've had occasions to experience this when (for one reason or another) I failed to take my usual dose of vitamin and mineral supplements. In each instance, my energy level and stamina were depleted. These experiences have reinforced the benefit of a good diet and supplements. I have found that self-discipline, proper exercise, and adequate nourishment have steadied the course of my journey with MG.

Being in touch with your body offers you choices of what is appropriate for you to eat. Through fine-tuning your awareness of yourself and your reaction to food over a period of time, you can achieve dietary habits and practices that become comfortable, natural, and conducive to health.

I urge you to take stock of your nutritional needs. *Getting old isn't a privilege given to everyone.*

POSTSCRIPT

A crème brûlée confession: Rebuilding my health was a priority for over a decade. I was disciplined with my medicine, exercise, and diet because it was important to my well-being. One day I allowed myself to make my favorite dessert, crème brûlée, and I ate the whole thing! My discipline went out the window and it felt good—for at least a few minutes. Fortunately, I don't do that every day, week or month, and I feel better when I follow a healthy diet (protein, fresh fruit and vegetables, and some starches). But, yes, I am human and I'm not perfect!

CHAPTER 7

THE QUEST FOR MUSCLE STRENGTH

The author Miguel De Cervantes created a lovable and unforgettable character in Don Quixote. The "Man from La Mancha" was intent on righting the wrongs of the world in very imaginative ways. We smile at his comic exploits and relate to his idealistic vision.

You may be thinking that this chapter heading seems to be another "impossible dream." After all, MG is a muscle weakness disorder. While it's difficult to exercise with a disease that causes muscle fatigue, it is important to regain strength. Can there be a viable program to strengthen depleted muscles? Well, that quest has been an ongoing project of mine for many years, and I'd like to share the fruits of my research and focus with you.

As I'm writing this chapter on muscle strength, there's a dramatic undercurrent in the air. At long last, spring has arrived in our Chicago area. This season has a special significance to those of us who have endured the freezing temperatures, snow, and ice of our extended winter. What a delight it is to smell the green grass, see the myriad colors of the flowers and the trees in bloom, and feel the warmth of a gentle breeze.

It is early morning and the birds are chirping as they forage for food and build their nests. This year the doves are nesting on top of our wood stack, the robins have chosen the roof eaves of our home, and the house wrens are occupying the birdhouses in our crab apple tree. I sense their frenzy as they prepare their nests for the next generation and listen to their songs with delight. As the day proceeds, the birds become quiet until the next morning. This bursting forth of life reminds me of my desire to build my muscle strength and the necessity of being physically active.

****

Despite the weakness inherent in myasthenia, I needed to find exercises that would boost my muscle strength. In the last chapter I explored the benefits of a nutritional plan. While a well-balanced diet and supplements are important, I realized that I also needed physical activity in order to build my muscle strength. At first I thought exercising would be easy. After all, I had attended fitness classes most of my life, played tennis and golf, and even (at one time) taught vigorous exercise classes. Surely building my muscle strength would not be that difficult. At least that is what I thought at the time.

As a member of a health club, I decided on an exercise routine. I planned to devote one day a week to fitness class, sauna, and steam room. I soon learned that energetic exercises depleted my leg and arm muscles, and the sauna and steam room sapped me even further. (Myasthenics become physically weaker with extreme heat.) After a few weeks, I tried again and the same thing happened. My whole body became limp after five minutes of vigorous exercise. It was very disappointing because I really needed to strengthen my muscles. Then my friend Marla suggested *yoga* and I signed up for a class. Fortunately, I met a wonderful yoga instructor, Judy Lipke, another person who changed my life for the better.

THE WHOLENESS OF YOGA

Although yoga began in India thousands of years ago, it was introduced in the United States at the Chicago World's Fair in 1893. Since then, classes devoted to yoga instruction have become a staple in our American culture. What is yoga? Yoga is the joining of the body, mind, and spirit in a disciplined form of exercise. It is a series of stretching movements and breathing exercises performed slowly and methodically. At first some postures may seem quite advanced and complicated for the average person. But there are simple yoga exercises that can be practiced by persons of all ages. It doesn't matter if you can't achieve the perfect pose. Patience, effort, and steady practice are essential.

My yoga teacher, Judy Lipke, was a bright, lively person with a great figure and a cheerful smile. She had taught yoga for more than twenty-four years, acquiring her Masters Teachers' Certificate at the Temple of Kriya. (Kriya=yoga of services.) She also was instructed by the world famous yoga teacher and author, Richard Hittleman, who adapted yoga postures to our Western culture.

To be honest, I was bored with yoga when I first started classes. I was accustomed to vigorous exercise with fast movements and repetition until it hurt. Now I was doing everything in slow motion and wondering how it could help. I am certain I would not have continued with the class had I been physically able to do any other kind of exercise. Judy told me later that she didn't think I would continue yoga since my mind was restless, my body impatient, and my emotions unsettled. Later she evaluated me as a "10" student because yoga helped me to discipline my energies to write this book and enabled me to harness my energy, especially learning to rest when needed without resentment. I'm grateful that I was persistent. I have learned that there is probably no other system of physical activity that exercises the whole body as does yoga. I also learned how important it is to keep your body active when you have a chronic disease or are recuperating from an illness.

THE MANY FACES OF YOGA

As I approached the challenge of yoga, I met many interesting people and a variety of instructors. One of my classes was filled with young mothers, all anxious to keep their bodies in shape. Their enthusiasm was contagious. Mona, the instructor for the class, was also young and passionate about yoga. She made yoga a fun and challenging activity, motivating us to exercise to the extent our bodies would allow.

In another class, one of the members was a priest who claimed that yoga was a great way for him to reduce the stress of his ministry. Other participants had obvious physical limitations. I clearly remember the elderly couple who arrived pushing their walkers and the middle-aged woman recovering from hip replacement surgery. All of them were able to take part in the class and work on rebuilding their strength. One yoga instructor actually conducted a class in a wheelchair, much to our amazement. She was a powerful motivator.

Actually, I had taken yoga classes most of my adult life before being diagnosed with MG. I had participated in advanced classes and accomplished most positions. I found it invigorating. Now, because of my health issues (shortness of breath, weakness in my arms and legs, and weakness in my spine) I went back to a beginner's class. Kathy was the instructor. She had taught yoga for over two decades and knew how to make a beginners' class challenging without fatigue. As Kathy talked us through each movement, I wondered how she knew exactly how that stretch or movement affected the different parts of my body. Sometimes I sensed what she said should be happening and other times I made slight adjustments in order to get the full benefit of the exercise. Everything was in slow motion enabling me to work on every muscle in my body to some extent. Kathy made sure each of us applied ourselves to the best of our ability. Yet she recognized that our bodies were all different and we should "listen" to the comfort level within us.

Yoga has helped me combat the fatigue related to myasthenia gravis. It calms and energizes me. It also strengthens the muscles in my arms and legs, and has enhanced my posture. After an hour of yoga class, my improvements are clearly noticeable. Other exercises—like aerobics and riding a stationary bicycle—exhausted my resources and left me feeling extremely tired.

HOW YOGA WORKS

Physical work is not the same as exercise. Physical work uses most parts of the body but leaves you tired. Yoga stimulates your body and leaves you invigorated.

Most forms of physical activity concentrate on developing a particular set of muscles. In yoga, however, every joint, muscle, tendon, nerve, and ligament, as well as all internal organs, are exercised, pressured, and stretched. Yoga provides the body with fresh, oxygen-saturated blood to revive and rejuvenate. It renews energy while calisthenics depletes energy. It also disciplines the body so the mind can expand.

There are many stages and types of yoga. Most beginners start with basic yoga breathing as well as standing and sitting postures. *Hatha yoga* teaches you how to quiet the mind by focusing on the breath and movement of the body. It relaxes, rejuvenates, tones, energizes, firms, and regulates body functions. *Astanga*

Vinyasa yoga is an invigorating and heat-producing practice following a sequence of breathing, along with a variety of seated forward bends, twists, and backbends. *Vinyasa yoga* utilizes many aspects of the Astanga Vinyasa system but it also explores different postures, breathing, and meditation. The primary goal of all the various types of yoga is to relieve the body of tension. It can also increase your endurance and flexibility for participation in more strenuous activities.

My first instructor, Judy Lipke, taught Hatha Yoga which deals with the physical and mental discipline of breathing techniques and exercise. Judy pointed out that yoga is neither a religion nor magic. It is hard work designed to help develop physical and mental competence. The exercises are progressive, adapting to one's capability, training the body and mind to work efficiently without effort. Yoga can relieve your minor aches and pains, increase your circulation, and even improve your sleep patterns.

After Judy retired, I continued Hatha yoga and also tried different types of yoga with other teachers. However, because of my health issues, I usually take a basic yoga class that concentrates on proper breathing, sitting, and standing—exercises that engage every muscle in my body. Occasionally, I'll test myself with a more aggressive class. I still enjoy a challenge.

An hour of yoga can change your life. You can perform yoga at so many different levels: a brief, relaxing interlude in a hectic life, a more demanding workout for strengthening and invigorating the body or a therapeutic practice for a physical ailment. Whatever your motivation or level of practice, yoga offers benefits that affect all aspects of your life: work, recreation, eating habits, family life, and relationships.

Yoga can be performed in some degree by any myasthenic who can move his or her limbs, most elderly people, and even those with disabilities. It takes desire and persistence. When I visited a senior citizens' yoga class one day, I was surprised to find they could all do more than I could when I first started yoga.

Regardless of age or health, I learned that through yoga I could exercise safely and succeed in rebuilding muscle tone and stamina. I discovered that yoga was nothing like calisthenics. I liked getting into postures slowly, and then holding a position as long as comfort permits. There's no forcing of positions, jerking or bouncing in order to "go further." You can determine how long you can maintain a position and you never have to compare yourself to anyone else in the class. However, as with any new exercise venture, I recommend that you check with your doctor before beginning a yoga program.

YOGA AND BREATHING

Breathing properly is probably the most important part of yoga. When I first got out of the hospital, I was weak, had little or no muscle tone, still had trouble speaking, and was tired most of the time. Had I thought about yoga then, I could have worked on breathing in a controlled fashion to work my lungs and strengthen my abdominal muscles.

Most people use their lungs at only one-third of their full capacity. Unlike most nutrients, oxygen cannot be stored in the body. Our bodies require a constant, steady, fresh intake of air. The first exercise in yoga (and it is an exercise) is to learn to breathe properly. Yoga breathing uses the respiratory system to its maximum potential, letting as much oxygen into the body while removing as much carbon dioxide as possible.

I was astonished by the number of young women in my class who, like me, did not know how to breathe correctly. Although I was reasonably agile and could do nearly all of the basic postures, I did everything wrong when it came to breathing. I would take quick shallow breaths or breathe only in the chest area. When practicing yoga, I tended to hold my breath, producing a state of tension instead of a relaxed posture. When I was asked to lie quietly so the instructor could show me how to begin breathing from the diaphragm, I would wiggle, squirm and even complain.

However, I was persistent and my instructors were patient. They helped me develop a breathing method that uses the respiratory system to its maximum potential. In yoga, this method is called *The Complete Breath*. We practiced the complete breath until it became second nature. Here are the instructions for the complete breath. I think that you'll find that by practicing this breathing exercise, you will feel less tired as your body functions more efficiently.

Sit or lie in a comfortable position. Straighten your back and thorax for easier breathing. Exhale slowly through your mouth emptying your lungs. Then inhale slowly through the nose, breathing deeply, consciously filling the lower part of the lungs with air by expanding the ribs and pushing the abdomen out. Your chest will lift. Continue the breath until the abdomen is fully extended; then expand the rib section outward to the sides and fill the mid-section of the chest with air. Finally let the breath fill the upper part of the lungs by lifting the chest area and letting it expand outward and to the side. Hold the breath for three to five seconds. With practice, you should be able to hold it for ten seconds or more to give the lungs a chance to absorb all the oxygen.

Now exhale slowly, gently contracting the lower abdomen first. As the lower lungs empty, the rib section will slowly deflate followed by the upper chest. This process of exhaling and then inhaling should take about five seconds.

Pause for a second or two before beginning the next inhalation. Repeat these steps three times and gradually increase your breathing exercise to ten repetitions. It is best to practice several times during the day so that this exercise becomes a part of you own breathing pattern.

There are many other breathing exercises, such as *The Calming Breath, The Cleansing Breath, The Cooling Breath, and The Breath of Sleep.* Many books have been written about yoga which describe the various methods of breathing in detail.

THE DIET FACTOR

I discovered an important side benefit of practicing yoga. In the last chapter, I detailed my nutritional plan and how it evolved. And I confessed that there were times when I had difficulty following a healthful food regimen. Then I found another advantage to practicing the discipline of yoga: it motivated me to adhere to my nutritional plan.

The yoga program recommended a diet emphasizing "natural" foods such as whole wheat flour, raw vegetables, brown rice, nuts, fruits, raw sugar or honey. This advice reinforced my nutritionist's guidance. It was similar to having duplicate and complementary pathways on the same journey. As the physical outlet of yoga exercise progressed, I became even more determined to persevere in the goals of my nutrition plan.

Yoga is helping to rebuild my strength. I am cautious, especially with my back. At the same time, I find that carefully planned yoga movements (used in a beginners' class) strengthen my back muscles. And I know that if I don't exercise, my muscles grow weaker.

As I continued my pursuit of strengthening my muscles, I looked for other paths to help my fatigued body and improve my mental function. Gentle yoga exercises and "the complete breath" were the initial breakthroughs of my search. (I had given up golf and tennis because of health issues.) Now I needed to explore other forms of being physically active. I found the answer in slow walking and isometrics.

THE TRAIL OF SLOW WALKING

When most people think of walking as an exercise, they think of setting a rapid pace. Fast walking is fine if you want a cardiovascular, stimulating activity. Slow walking is calming and focuses on stretching muscles to relax them. I discovered that slow walking increases my energy and improves my sense of well-being. Of course, I am careful not to overdo because that would lead to fatigue. I also remember to breathe properly when I walk. (Yoga taught me breathing should be from the abdomen.) Slow walking is an exercise most people can do. No special equipment is needed other than socks that fit and walking shoes which provide good support.

The intangible in slow walking is the discipline of persistence. A good example of a dedicated walker was my friend, Helen Williams. Helen was also a myasthenic. A mutual acquaintance introduced us and we became close friends. For over a decade we saw one another for coffee, lunch, and long visits. We talked about our families and how a positive attitude makes our day.

When I produced talk shows for a local TV station, I asked Helen to be our guest, along with Dr. David Richman from the University of Chicago, to discuss myasthenia gravis. While Dr. Richman informed us about medical treatments and new medicines, Helen presented the human perspective of an MG patient.

Years later, when Helen was diagnosed with cancer, she had a party to celebrate her life. I met her husband, Walter, her daughter Geri, her sons Stephen and Walter III, and many relatives and friends from all walks of life. It was an awesome party filled with stories of how Helen had touched the lives of so many.

When death took Helen, it was as if an angel had come and lifted her soul. Once again, we gathered to celebrate her life—this time at her church in Maywood, Illinois. Helen knew I was writing this book and wanted to be a part of it. Fortunately, Helen left me a story of her own experience with MG which I am happy to share with you. I hope it encourages you to be like Helen in dealing with your mystery guest.

A WINNER NEVER QUITS—A QUITTER NEVER WINS

By Helen E. Williams

I was diagnosed with Myasthenia Gravis in May, 1980. My story is about exercise and optimism. Thinking I would be able to actively participate in a fitness program, I joined the Gottlieb Fitness Center. Last year I received my ten-year tee shirt that said, "I'm an original."

I started walking at our local high school track. I walk at the Center in the winter. When you have MG, walking a distance is not an easy feat. I set a goal of one mile daily. I walk a half mile, then retreat to my car and rest at least fifteen minutes and then finish the last half mile. I limp because I have severe arthritis in my knees, legs, and feet. Also, my left foot does not work as a team with my right foot. On occasion it will go east when my body is going west.

The other walkers at the Center whiz by me several times, walking briskly and leaving me in the dust. Yet some days they don't show up. I am consistent. There are some days I am unable to walk but I still try.

I met a pastor on the track. He had two massive coronary heart attacks and was very sick yet he showed up daily. He was impressed with my persistence and told his congregation about me. They invited me to speak at their church. I told them about walking and how I tried to inspire others to walk by showing up every day. In the face of adversity and insurmountable odds, I keep trying.

My co-walkers have a special respect for me and have been very kind. When I waiver and look like I won't make it, they pause or wait in their car until I come off the field. This is an act of kindness.

I know today that in life, "The race is not given to the swift, but to those that endure to the end." God has been good to me. I'm happy to be one of God's Christians. I use my walk time to meditate and pray and thank God for being alive. I always feel mentally filled."

Helen endured to the end. She was a winner; she was my friend.

****

Dealing with disabilities with optimism and persistence is a challenge. Another friend, Bob Graybill, had a number of physical limitations after cancer, brain surgery, and a stroke that left his left side partially paralyzed. He, too, became an inspiration.

Bob was an avid golfer with a twelve handicap. (For non-golfers, a twelve handicap is a low number consistent with a very good amateur player.) After his illnesses, he was determined to play golf once again. Walking was his first issue. He worked with a therapist until he could "tell" his left leg to move. He would ask, "Is it moving?" He could not feel the movement but was able to send a signal to his leg to move forward. Bob's left arm was also affected and he worked daily to force movement in his arm.

My husband enjoyed playing golf with Bob. Jack would work the golf glove on Bob's left hand and off they would go in the golf cart. Jack shortened a fairway wood with an oversized grip which Bob was able to swing predominately with his right arm. Bob could hit this club about 120 yards, usually straight down the middle of the fairway. His fortitude and persistence were truly phenomenal. Regardless of his golf score, Bob was a winner.

Bob's endurance and Helen's story remind me that behind every dark happening and difficulty, there is a hidden blessing if we follow the light in our heart. Do not regret the past. Move forward and be thankful. Gratitude can cast out fear.

Flowers open slowly to perfection when they are in the sun. You can do the same with patience, a quality we all want and most of us want it *now*. If we can endure and persevere, we too can be "winners that never quit."

THE WAY OF ISOMETRICS

Anyone who has been bedridden for more than a week knows that muscles waste away without use. Regaining muscle strength is a slow process but it can be done. In my own journey for rebuilding muscle tone and strength, I sought another form of exercise in addition to yoga and slow walking. I discovered *isometrics*—a simple type of exercise which is the basis for yoga and Pilates.

Isometrics are exercises that involve contracting muscles by pushing, pressing or pulling against a resistance such as a door frame or other muscles. The word *isometrics* consists of *"iso"* meaning same and *"metric"* meaning size or distance. In isometrics, there is no shortening or lengthening of muscle. Because isometrics strengthen muscle tissue on isolated muscles without undue strain or movement throughout the rest of the body, it is especially helpful for people with disabilities or those who have a limited range of motion.

I started with simple isometric exercises such as pressing the palms of my hands together, holding the contraction for several seconds and then relaxing. This works the muscles in my arms, upper back, and shoulders. While sitting at the computer, I draw in my abdominal muscles and repeat the pattern of contraction, holding, and relaxing. I also use the doorway frame as an immovable object of resistance. I raise my right arm above my head and press my arm against the door frame, alternating with my left arm.

141

Again I repeat the pattern of contraction, hold, and relax.

Most physical therapists advise holding the contraction for five seconds, relaxing, and then repeating the muscle flex ten times. Gradually I was able to perform these and other isometric exercises with two sets of ten repetitions. There are many fine books on isometric exercises that describe specific contractions in detail. As always, I recommend that you request your physician's approval before undertaking isometrics or any other type of exercise.

Most of us—myself included—sometimes use the excuse that we don't have time to exercise. *...It's raining, I can't go walking today. ...Too bad, I have a meeting and I can't make my yoga class.* Well, when it comes to isometric exercises, the excuses slip away. These exercises can be done while waiting in line at a grocery store, watching TV or lying in bed. No special equipment is needed, only a desire to strengthen muscles in a straightforward fashion.

The benefits of isometric exercises are numerous. These workouts burn calories, strengthen and tone muscles, improve the ability to hold contractions, and boost energy levels. It is amazing that something as simple as isometrics can accomplish so many positive results.

****

Don Quioxte relied on his faithful friend Sancho Panza to assist him on his creative journey of doing good deeds in the world. The adventures of my life have been simpler. Many fine professionals have aided me in my quest for rebuilding my health. Over the years, my nutritional plan and muscle strengthening exercises of yoga, yoga breathing, slow walking, and isometrics have gradually become a way of life for me. Yet there have been times when my discipline has wavered. When this happens, I lift myself up and begin again. I'm grateful that I've persisted in searching for ways to restore my well-being in positive ways. It has not been an "impossible dream" because of the love and support of my family, and the abundant graces of a loving God.

POSTSCRIPT

Growing up on a farm was a wonderful experience in healthy living. Activity and work were part of our natural day-to-day existence. As a young girl, I remember climbing trees, riding horses, learning to swim in a nearby lake, and sharing a bicycle with my brothers and sister. Every day included physical exercise. There was little thought of fitness centers, weight lifting, jogging or long-distance bicycle trips. We didn't have to seek ways to keep fit.

I was ten years old when my family left the farm. Over time my forms of exercise evolved to match my lifestyle. I urge you to find your own activities to enjoy and keep yourself well. There is no mystery to the benefits of exercise.

CHAPTER 8

A BLESSING IN DISGUISE?

It has been a long journey. Sometimes the path was straightforward but at times there were road blocks and detours. My faith has guided and kept me on the right course. No matter what happens, I feel secure in the firm knowledge that God is with me, guiding me through the twists and turns of my passage.

I recall the inspirational poem "Footprints in the Sand" written by Margaret Fishback Powers. In her classic poem, Ms. Powers relates the story of a person, weighed down with difficulties, trudging through the beach leaving one set of footprints in the sand. Where is God during these desolate times? Has God abandoned this suffering person? God answers that He was there all the time, carrying the person in crisis. The single set of "footprints in the sand" belongs to the Lord. I can relate to that beautiful poem because I believe God has been with me every step of the way.

To some, faith or trust in God is something for which there is no proof. By the grace of God, I have a strong conviction that has been forged in the fire of difficulties. At times it has seemed that circumstances were working against me, but I know that with God's help I will prevail if I persist in my belief.

My faith is like the rainbow that promises calmness and beauty after the storm. Throughout times of trial, my faith has given me an inner strength and an assurance that God will sustain me.

Life can be rich and meaningful when I allow the Lord to guide my decisions. In the previous chapters I discussed the importance of nutrition and exercise in rebuilding my strength and energy. As a result of research and consultation with professionals, I found a nutritional plan and exercise regime to promote my recovery. Yet I knew that I also needed to work on developing my spiritual life in order to face and deal with the challenges of a chronic illness.

Each of us encounters God in different ways. I find God by reading His Word in the Bible and worshipping with a community of believers at church. At times I have been able to attend a day of retreat, time solely devoted to being open to God in thought and prayer. But God is not confined to the sanctuaries of churches, synagogues or mosques. He can be found in the myriad stars in a heavenly sky, the vastness of the oceans, and the beauty of nature.

I am especially drawn to communing with God in the quiet and serenity of our home garden. In the spring, summer, and early fall, I spend time in the garden delighting in the flowering plants and observing the many birds attracted to our feeders. Here I can place myself in God's hands and enjoy the wonder of His creation in our suburban oasis. It is a sacred time when I grow spiritually in my relationship to God. During these occasions, I thank God for all that He has done for me.

Was my mystery guest a blessing in disguise? Certainly it has changed my lifestyle. Sometimes it is frustrating because myasthenia gravis is so inconsistent and unpredictable—qualities that are diametrically opposed to my organized personality. Still my patience is growing, and I am more aware of touches of grace in my surroundings.

ANGELS IN OUR MIDST

Have you ever wondered if you have a guardian angel, someone who watches over you? In my Sunday school class, we learned that angels were spiritual beings without earthly bodies, messengers of God. The Bible records many instances of God sending an angel (in a person's vision or dream) to give a special work or mission. If angels exist in our world, and I believe they do, I think their presence is reassuring evidence of a loving God who is forever watching over us.

Angels keep appearing in my life. As my children were growing up there were many "angel moments." I recall when they were small, one of their favorite things to do after a snowfall, other than build a snowman, was to "make" an angel in the snow. They would lie in the snow on their backs. They worked their legs and arms in and out in a manner to form what looked like an angel in the snow. They loved the snow. To them a snowflake was light, fluffy, and pretty. It wasn't until they were older and had to shovel snow that they took a different view. Pretty yes, but when you are shoveling twelve inches of snow, it is not light and fluffy.

Susan also loved playing an angel in the church play. We made her wings from a white sheet and wire. Instead of an angelic smile, she had a mischievous grin. It was a happy memory.

There were also times when I felt there were guardian angels watching over and protecting my children and our family. Sometimes it happened during moments of need, other times their heavenly influence occurred in our everyday life. Little touches of grace when a child was sick or a near accident averted. Then there were the growing pains of choosing friends or schools or activities.

Nothing happened that was as spectacular as the burning bush Moses experienced in the Book of Exodus. Yet looking back on many everyday events makes me realize we were often gently guided. Perhaps it was our guardian angels working overtime, especially when our children were teenagers.

Then there were unusual signs in our lives. An angel brushed against us when my father-in-law died. The church was filled with family and friends. At the gravesite, a smaller group of family and friends (including his dove-hunting buddies) gathered. As the service concluded, a large flock of birds flew up from a nearby tree into the sky. They were doves. It was an amazing coincidence.

Angels have been in my life in simple ways—a smile from a friend or a stranger at just the right moment. Sometimes the angelic flutter appeared in the form of a bird chirping, a butterfly appearing at my window or the whisper of the wind. Whatever my perception of angels may be, I know there is divine assistance available to me at all times. When I am aware of their presence, I am at peace. Sometimes I think an angel becomes an actual guardian on my behalf, keeping me safe and helping me to make the right choices.

Then there are the other kinds of angels. Reflect on the people you know who have changed your life in some way for the better. Think of the child, friend, teacher or neighbor who by their presence created an atmosphere of love or were there for you at just the right moment. Is it so difficult to believe God sent them on a mission to you?

With all the blessings God has given me, I feel a certain sense of urgency to share the experiences of my life. I hope you, the reader, will find some comfort and encouragement in the following sections in dealing with your chronic illness. These suggestions mirror my own philosophy of life. They have guided me through the highs and lows of dealing with myasthenia gravis.

ASSESSING AND MAKING CHOICES

Life does indeed give us "mountains to climb" and challenges to overcome. If this happens to you, take time to evaluate yourself and assess your goals. Whatever your goals are, concentrate on reaching them one step at a time. Take stock of your life. How do you spend your time? What are your priorities? What is important to you and your loved ones? What is ego-enhancing?

Concentrate on what is worth "being" instead of what is worth "having." Stop measuring your life in quantities, number of committees, friends, and accomplishments. Think in terms of quality and rid yourself of trivial obligations. Give up trying to be a "super" person who has control of everything at home or at the office, leads in community affairs, and raises *perfect* children. ...Forget perfection.

Having a general daily regimen helps to give direction and order to activities. However, don't be too rigid with plans. Always leave them flexible enough to allow for unexpected and spontaneous joys. Slow down; take time to "smell the roses." Don't waste your anger on small matters. Everyone has a "mountain to climb," some are just bigger mountains than others.

How can you put some of my experiences to work in your life? First, think positive, and then...be persistent. Think about the old story of the glass half filled with water. The pessimist sees such a glass as half empty, and the optimist perceives it as half full. Then imagine you are like that glass and the water represents your health. Are you half full of health? Or are you half empty?

I am basically an optimist. I believe counting my many blessings is better than worrying about problems, especially when it concerns my health. Pessimism is not healthy. We can think of ourselves as healthy or sick; the choice is up to us. Because I live my life thinking my glass is half full doesn't mean that I don't have moments of questioning, complaining, and making the wrong turn in my passage.

It's not that I don't want to share my health issues and the difficult journey I have traveled, but I don't want to dwell on adversity when every defeat and heartbreak has its own lesson. I do not expect to walk through the remainder of my life on mountain peaks. No matter how hard I try or how much I persist, there will be days, weeks, months or maybe years when everything I attempt is frustration and sometimes failure. Yet I continue to think of my glass as half full. My prayer is, "In failure, preserve my faith. In success, keep me humble."

REGENERATION & HEALTHY HABITS

We have more than a choice between optimism and pessimism. We can choose to worry about sickness or use positive ways to keep our "glasses full." Both ways are more than metaphors. If we decide on the latter route of affirmation, focusing on feeling better and living life to the fullest, there are ways to reach our goals.

One optimistic way to keep our glass full is *prevention.* This means living our lives so that we're on our guard to protect and maintain our health. Not much of the water falls out of the glass in this mode of operation. Another positive way of dealing with your health is *regeneration.* Take sleep, for example. Sleep is regenerative. When you are tired, you sleep and then awaken refreshed and regenerated. Sleep is a natural process of self-renewal that happens every day.

Regeneration—the way I'm using that word—means to become formed again, to use the power within to "create" health. Exercise is regenerative. Imagine sitting indoors on a nice day watching television, feeling tired. Then you get up, turn off the TV and go for a walk. For the first hundred yards or so your feet feel heavy. Then your muscles begin to warm up and your heart begins to beat a little faster. Gradually, you become energized. You don't feel tired any more. You have begun to regenerate and your "glass" fills with a little more water. That night you will probably sleep better.

If you are overweight, you may find that reducing is regenerative. Too much weight can increase your vulnerability to disease. But thinking about future sickness is a pessimistic way to visualize your personal "water glass." If you lose a few pounds, you will feel better. Losing weight will also help your heart, lungs, and even your feet carry you through your day more easily. You may say, "It may be easy for some people to lose weight but it's hard for others." While this may be true, I can guarantee that if you cut back the amount you eat (one small 1/2 cup serving instead of one or two large servings) and if you watch the number of calories and sugar intake, you will lose weight. Until your body adjusts to the change, eating less is difficult but it will gradually get easier.

Good diet is literally a cornucopia, an abundance of regenerative possibilities. The food we consume is our greatest exposure to our environment. When we eat, we are definitely making our glass of health either half empty or half full. But eating is so complicated and diverse. What simple way is there to know how to eat in a way that regenerates our bodies? How can we move beyond worrying about sickness and even beyond the idea of prevention?

A friend answered these questions when she said, "Health is not about itself, it is about everything." Real health is more than being sick or not sick, more than preventing illness. Health is everything that fills our glass. Health has to do with where we live, what jobs we do, how we spend our free time, and how we feel about our personal sense of independence.

How do we apply these thoughts to our eating patterns or habits? Obviously, it helps to think about what you eat. If good food is the best way to regenerate ourselves and refill our glass, and if health is truly about everything, then eating requires a variety of healthy foods.

When we're speaking of health regeneration, let's keep in mind that we are usually healthier than we realize, and our potential for improving our health is almost limitless. If you doubt that concept, just think of all the diseases and illnesses you don't have and probably never will due to your body's resistance. You can use that knowledge of your personal wellness in very practical ways. What you are doing when you "put more water into your glass" is enlarging your inner islands of health.

What about toxic substances? Drinking too much alcohol can cause fatty cells to form in your liver which can lead to cirrhosis. If you do drink alcohol, it's vital to moderate your drinking and thus preserve your body's ability to renew itself. In fact, some studies show that small amounts of alcohol in the diet can actually be beneficial. No such positive claim can be made for tobacco, the other common toxic substance. Research continues to reveal all the health negatives of smoking tobacco. Your health can improve if you knock down the wall that surrounds these toxic substances. And I have found that establishing healthful habits is the best way to circumvent the negatives of toxic substances.

Sometimes it's helpful to outline a plan of attack when it comes to developing healthy habits. Here are some of the habits I've tried to develop in my own life plan:

1) A good diet including the regular use of food supplements to enhance and bolster my unique needs. My well-balanced diet should also help me maintain my weight at a reasonable level. Years ago, infections and malnutrition were the number one killers of people. By learning how nutrition works, you can take control of your quantity and quality of life.

2) Exercising at least three times a week with activity vigorous enough to breathe deeply, raise my pulse and perhaps even "break into a sweat." For me, yoga promotes emotional strength by teaching me to concentrate and remove distractions from my mind. It teaches me to act from awareness rather than emotionally. When I could no longer play golf

and tennis, I added slow walking and isometrics to improve my circulation and boost my muscle strength.

3) Reducing the use of toxins in the environment. Drink alcohol in moderation or eliminate it completely. Avoid tobacco and second-hand smoke. Drink purified water.

4) Take life one day at a time. Turning my attention toward the most obvious and apparent good in my life, I can't help but notice I am always grateful for something. My gratitude is the key that allows me to recognize even more of the good that is all around me. I have learned that when a door closes, another opens. And I have learned that life is shaped by my thoughts. In order to change my life, I must change the way I think. The first step is to acknowledge that change is needed.

5) Relax and schedule rest periods every day. Learning to put more regenerative relaxation into my life involves resting, being with friends, and finding ways to reduce stress. If you can identify the source, you can look for ways to manage the stress in your life.

6) Seek a quality of life. When your health fails, decisions become more difficult to make. A sense of helplessness is normal as your priorities in life change. Rather than calling it a barrier, consider it a bridge or perhaps your "mountain to climb."

7) Emotions are energy in motion, strong feelings expressed outwardly or inwardly. They are the

motivating force in my life. This motivation can be productive, empowering, and creative or it can be negative and abusive. I believe that whether we are happy or sad depends on our ability to manage our emotions. Sadness is a place. Some people live there too long. Think in terms of *this too shall pass.*

8) Keep your glass half full. Cultivate positive feelings. There is a biblical saying, "Whatever a (person) thinketh in his heart, so is he." Your heart connects with the feeling side of you. Positive feelings open up pathways to your soul. Cultivate positive feelings and use them as your internal compass to attain the life you desire.

9) Educate yourself with information about your disease. Find the doctor who has knowledge about current treatments. While I believe proper diet can help heal my body, I know diet alone cannot make me well. Proper medical treatment is needed to aid the natural healing of my body. Knowing more about my disease will calm my anxiety and fears. Learn all you can and rely on your doctors.

10) Be prepared—carry a medical ID card in your wallet. List all medications as well as your allergies to medications. List vitamin and mineral supplements you are taking. Avoid hot and cold weather extremes that may exaggerate weakness. Wear medical identification jewelry that is recognized by health professionals. This

can be obtained from "Medic Alert," 1-800-432-5378 or **www.medicalert.com.**

In summary, learn to think beyond sickness. Do less worrying about your ailments and count your health blessings. Make a start and you will actually feel your "glass refilling." I know that I am at high risk if I don't get my rest, stay on my diet and exercise programs, and take my medication. It would not be fair to my family or myself to knowingly do something to cause an MG crisis or fail to do more each day.

Life is not static. In the last twenty years, I have had some additional trials to my health and well-being. People with chronic illnesses often have to struggle with other difficulties. Again, I relate how I dealt with my new challenges in the next section to give you hope and courage for your own journey.

NEW HEALTH CHALLENGES

In 1986, I had an appendectomy and some additional exploratory surgery. The surgeon was surprised to find that I was basically a very healthy woman even though I had been taking prednisone for seven years. In fact, I was given additional large doses of prednisone before the operation to offset the risk of a myasthenic crisis during surgery. I credit my good report, despite the known serious side effects of prednisone, to my rigid nutritional program of a good diet and vitamin supplements.

A Blessing in Disguise?

In 2001, I was in London with my husband when my eyes began to spasm and my balance was uncertain. We flew home the next day and my doctor ordered an MRI which revealed a brain tumor. A team of surgeons removed a non-malignant meningioma. Two cranial nerves were stretched, causing residual damage to cranial nerves X and XI. My right vocal cord was paralyzed and I could not raise my right arm shoulder high. After range of motion therapy for four months, I was able to use my right arm almost normally although there are some limitations in my yoga exercises. I had had twenty-four years of adjusting and compromising with my body but pain was never a part of MG. Since my brain surgery and cranial nerve damage, I have had to deal with pain on an everyday basis.

A prosthesis to modulate my vocal cord along with months of voice therapy have been helpful. I have a number of what I call "nuisances" regarding the vagus nerve but I am fortunate it was not more destructive. Once again, I am learning to live with a new problem or, as I like to say, a new challenge. I call it my St. Helena's eruption—a whirlwind of difficult experiences and new issues on my agenda. On the lighter side, I can no longer use my voice to sing. However, my husband (half teasingly?) reminds me that my musical "gift" has not changed; my talent lies in listening!

As a myasthenic, the anesthesia used during surgery has to be carefully regulated. Preceding the surgery, it was necessary to take heavy doses of prednisone in preparation for the eight to nine hour brain operation. Unfortunately, this strong dosage precipitated a mild myasthenic crisis which took several months to resolve. My bone density dropped considerably and my osteoporosis worsened.

Other difficulties developed the following year. In the fall of 2002, I woke up one morning and could not get out of bed. I soon learned I had fractured my lumbar spine. Once again, I was faced with many months of rest and rehabilitation. Thankfully, the fracture has healed well. Despite these reversals, I believe that the positive things I have outlined in this book have helped me to cope.

Since the fracture of my lumbar spine, I have been advised not to play golf or tennis. Now yoga has become my main form of exercise and I take classes two to three times a week. I also do slow walking and isometrics. Today, my body is stronger and healthier. Also, with the help of a friend and co-author, I have been given another chance to complete this manuscript.

Presently I can (with proper rest) do most things in moderation, at least for short periods of time. I refuse to rush. I take little periods of rest, exercise, take vitamin and mineral supplements, and watch my diet. The main difference between what I did before MG and the present is that I maintain some control of my life. Now I manage my schedule instead of letting it run me.

LESSONS LEARNED

Each person experiences fear in different ways and the manifestation of fear may appear in diverse forms. Death is not the enemy; living in constant fear of mortality is the true enemy. I will continue to live and think as actively, creatively, and positively as I can, knowing that longevity by itself can be sterile. Vital feelings and thoughts give meaning and depth to life and provide a sense of the possibilities of human existence. What some people find impossible, I call a challenge.

I have faith in my God-given abilities. I look to the future fearlessly, living one day at a time. The wise words of the Bible suggest, "Do not be anxious about tomorrow, today's own troubles are sufficient for the day." When I begin and end each day with the proper attitude of mind, I live a better life. Sadly, some people never enjoy the present moment because of the regrets of yesterday or the fears of tomorrow.

Keep in mind that the circumstances of each day are always changing. Today's world and what I do today will be different from what happens the following day. Perhaps tomorrow will be the day I encounter a new person, an idea or an experience that will drastically alter all that has gone before. Today's loneliness may be the prelude to tomorrow's shared pleasure—a delight made all the sweeter by the bitterness of the experience that preceded it. Remember that darkness passes but what one learns in the darkness remains with you forever.

Once I was an organized and predictable person who liked to be in control of situations. Now I have learned the meaning of the phrase, "Let go, Let GOD." For me, this saying signifies freeing myself of worries, fears, and anxieties while I place my confident trust in the Lord. Before I begin the activities of my day, I make time for God. Every day is a new beginning. My attention is placed in the presence of God. I unburden my heart and mind of any fear, worry or doubt. As I release them to God, I picture them floating away. And in the spirit of faith that accompanies my "letting go and letting God," I claim peace, love, and inner joy.

I believe, "God gives us talents—that's God's gift to us. What we do with our talents is our gift to God." With God's help, my future looks brighter. With my doctors' help, I am on a positive medical approach to cope with myasthenia gravis. My physicians' thinking is progressive and current and I feel confident that I am in good hands. Their professional support combined with my diet, exercise, vitamin supplement program, and plenty of rest help me to lead a productive and normal life.

When I use the phrase "normal life," I don't mean one that is without limitations. While I don't do as much as I did before MG, I belong to several organizations, entertain, travel, participate in a yoga class, and enjoy my five grandchildren. I spend many hours doing research and writing because I want to encourage others to work at maintaining or restoring their health. What I do to help others with a mystery guest such as mine is my way of giving a return for all the blessings in my life. It is my gift to a loving God.

Placing myself in God's hands has allowed me to overcome negative habits and challenging situations. Seeing the good in my life makes me realize how grateful I am. My life is joyful not because of what is happening in it, but because of the person I have become. I choose to see things with an optimistic attitude. Like a hopeful child, I have a natural tendency to expect positive outcomes. I have a passion for life that far exceeds any difficulty I have ever known. And peace flows from the approach I bring to my life.

Over the course of my illness, I have made an important discovery about life. There is a way to transform time into energy. The way we choose to live and the depth of our feelings, our ability to love and be loved and take in all the colors of the world around us, determine the worth and true extent of whatever time we have. If you can't enjoy the moment, you may be a burden to those around you.

I heartily relish life because I've learned how to minimize the effects of MG. While I'm taking my medicine and following a healthy regimen of exercise, rest, and diet, I know that I'm doing all that I can. I'm also aware that new and better drugs are on the horizon. Progress in research is being made every day, so there is hope for the future. Perhaps my mystery guest will succumb to some new medication or therapy. In the meantime, there is always the possibility of a remission.

Vince Lombardi, the former legendary coach of the Green Bay Packers, said that "If you think you can, you can." But thinking you can is only the first step. You must also persist. Where health is concerned, perseverance can mean the difference between getting well or not. My perception of myself and reality remains on a positive note. (I have zero tolerance for negative thinking.) I continue to strive toward real and permanent growth, and refuse to be discouraged by any delays. I am steadfast.

Life does indeed give us "mountains to climb" and obstacles to overcome. I have faith that through the power of God in me I am more than equal to the challenge. I see a beautiful world, hear its music and poetry, smell the fragrance of each day, and savor the moment. In my enthusiasm for life, I hold certain truths close to my heart:

Life is difficult....God is merciful...Heaven is sure.

So we return to the initial question posed in the title of this chapter. Is myasthenia gravis a blessing in disguise in my life?

I do know I have learned to relax and take time to enjoy the simple things in life. Things that once seemed meaningless have taken on greater importance; things that were once significant seem inconsequential. My life will never be the way it was before I was diagnosed but in some ways it is better. I have a larger purpose in my life—to help others.

My choices and actions help to bring harmony to my day at home, with friends and even a brief encounter with a stranger. Just as musicians take their signals from a conductor, others may take their cues from me. A smile tends to bring smiles in return and kindness helps to heal wounded hearts. So, yes, my mystery guest is indeed a *blessing*.

In my journey there have been road blocks, detours, and mountains to climb. I appreciate the struggle of climbing a "mountain" of difficulty. It's as visual to me as rainbows, clouds, stars, and sunsets. My "mountain," my uninvited mystery guest was (and is) a neuromuscular, autoimmune disease called myasthenia gravis.

I hope my thoughts have led you to grapple with your own "mystery guest" in faith. In the future, I envision a world in which spiritual seekers are nourished with divine strength and experience inner peace. Yesterday is over; tomorrow is yet to come. Let us live in the present moment, savoring its joys and learning from its challenges. Then, with God's help, we will endure and *climb the mountain* for I know there will be a magnificent view from the top.

POSTSCRIPT

A favorite Bible verse of mine is in the Book of Micah 6:8—

> *He has showed you O man what is good. And what does the Lord require of you? To act justly and to love mercy and walk humbly with your God.*

PART II

ANALYZING THE MYSTERY

CHAPTER 9

INVESTIGATING HOW MG WORKS

When I was a little girl, I loved to watch my mother in the kitchen as she prepared meals for our family. What a treat it was when I learned to make some simple recipes and listen to the oohs and aahs of my usually appreciative (and built-in) audience. My enjoyment of cooking and baking has continued, although my understanding and perception of the process has changed over time. Now I like to know the *why* of cooking and baking. It's not enough to follow the recipe exactly, watch the process unfold, and enjoy the results.

Something similar happened to me after I was first diagnosed with myasthenia gravis. I needed to know the "why" and the "how." You may have felt the same way, searching out books, magazines, the internet and other resources for answers. How does MG work? What are antibodies and how do they function in an autoimmune disorder? I had lots of questions. What I discovered is that searching for and gaining knowledge about my illness helped me to cope with my new situation. After all, something's happening within our bodies and we should try to understand it as best we can. Knowledge is power.

Perhaps I needed to exercise a certain control over what was happening to me. Or maybe after the initial shock of diagnosis, my mind was hungry for knowing as much as possible. If you or your loved one has MG or some other chronic illness, I hope this section will shed some light and give you an idea of how to proceed. I'll try to unmask the mystery of MG in simple terms.

DEFINING MG

Let's begin with our definition of myasthenia gravis and then break down each part of its meaning. MG is a chronic, autoimmune, neuromuscular disease which involves varying degrees of muscle weakness. If we look more closely at the name of the disease itself, we find that the words *myasthenia gravis* are derived from Latin and Greek words meaning "grave or severe muscle weakness."

With proper diagnosis and treatment, most cases of MG are not in the grave category. However, if the breathing muscles are affected or compromised, the myasthenic's condition can be potentially quite serious and even life-threatening.

Chronic describes the continual, unremitting nature of myasthenia gravis which is incurable. Once it is diagnosed, it can be treated and controlled. But it does not "go away" or subside over time. This was a hard concept for me to grasp and eventually accept. Of course, there is always the hope that medical research will find a cure. But at the present time, MG is an incurable, chronic condition. This harsh reality is all the more reason for exploring the "whys," seeking the finest professional treatment, and learning from others with the disease.

Myasthenia affects voluntary or skeletal muscles, the muscles controlled by impulses from the brain. These are the muscles used in opening and closing one's eyelids, chewing, swallowing, smiling, holding one's head erect, moving arms and legs, and breathing.

Often the first indication of the disease is in the head and neck muscles. You may wonder why these muscles are a favored target of MG symptoms. The muscles in the head and neck contain the *smallest* concentration of acetylcholine receptors vital for muscle contraction. When there is a breakdown in the transmission of acetylcholine from nerves to muscles, the head and neck muscles (with fewer receptors) are more vulnerable. (Fortunately, the smooth muscles of the heart, digestive tract, and other internal organs are not compromised by myasthenia.)

Ironically, MG is neither a disease of nerves nor muscles. The problem lies in the "gap" between nerves and muscles, the neuromuscular junction—where nerve signals from the brain to the nerves are transmitted to muscles. The syndicated medical columnist, Dr. Paul Donohue, describes the distinction between a normal nerve-muscle activity and the "glitch" in myasthenia gravis:

"Nerves release a chemical message called acetylcholine (ACh). The acetylcholine has to swim a small gap between nerve and muscle and land at a docking station. When (ACh) arrives at the docking station, a muscle contracts. In myasthenia, the receptor docking stations are blocked with antibodies. Acetylcholine cannot activate the muscle properly. The result is a feeble muscle contraction."

I like Dr. Donohue's analogy of a chemical messenger (ACh) floating between nerves and docking stations on the muscles. It gives me a mental image of what is happening. Nerve cells in the brain release the chemical messenger—ACh—which is transmitted from nerves to muscles through the neuromuscular junction. Normally, the ACh attaches to the muscle receptors and the muscles fibers contract. The strength of the contraction depends on the amount of ACh molecules the muscle receives.

If the acetylcholine "docking stations" or receptor sites are blocked, changed or destroyed, the muscle cannot contract normally. This is the situation in people with myasthenia gravis. Their immune system produces antibodies that attach to the receptors on muscle tissue, preventing ACh from docking.

When muscles fail to get steady spurts of ACh, they cannot maintain a contraction. If the eyelid muscles are involved, the eyelids droop and the myasthenic has trouble keeping his or her eyes open. When muscles of the throat are affected, the patient may have difficulty swallowing food. The most serious dilemma arises when breathing muscles become impaired during a myasthenic crisis. In all of these cases, the muscle fatigue or loss of muscle control is due to the lack of transmitting the necessary signal in the nerve/muscle gap. And the culprits in the blockage of "docking stations" are *antibodies* produced by an immune system which has gone haywire.

THE ROLE OF ANTIBODIES IN MG

Antibodies are proteins that help us ward off harmful invaders, infectious agents such as bacteria and viruses. Generally, antibodies protect our bodies against diseases. When we get a virus or a bacterial infection, our own antibodies attack these foreign invaders and help us fight the illness.

In the case of MG, antibodies destroy or block those all important receptor sites, the "docking stations" on muscle tissue. For this reason, the acetylcholine (ACh) messenger cannot attach itself to the muscle and stimulate the chemical reactions needed for a normal contraction. Because a myasthenic's own antibodies are the culprits, MG is classified as an *autoimmune* disease. Ironic isn't it that our own bodies wage warfare on ourselves? You may be familiar with other autoimmune diseases including rheumatoid arthritis, multiple sclerosis, lupus, type I diabetes, and ALS.

(You may be wondering what happens to the acetylcholine once it stimulates muscles to contract. Our bodies also produce a special enzyme called acetylcholinesterase which then breaks down the acetylcholine. This breakdown prevents any further stimulation of muscles to contract.)

Let's begin with *why* the ACh receptor sites are deficient. This is central to understanding MG. People with myasthenia gravis produce antibodies which destroy receptor sites faster than they can be replaced. This decreased amount of receptor sites results in muscle weakness. In myasthenics, a small gland in the upper chest, the thymus gland, acts like a command center for an immune response. It activates cells to fight "known" invaders of bacteria and viruses. But it also stimulates the production of antibodies that destroy "normal" cells.

In the case of myasthenia gravis, antibodies attack the ACh receptor sites so necessary for muscle contraction. You may have seen these antibodies referred to as AChR—acetylcholine receptor antibodies. This is a short way of describing antibodies that block or destroy the muscle receptors for that all-important nerve messenger, acetylcholine.

Research science has unearthed much of the "mystery" of how myasthenia gravis works. In fact, MG is one of the best understood of all the many autoimmune diseases. Yet many questions still remain. What is the exact relationship between the thymus gland and myasthenia gravis?

Early in life, the thymus gland helps in the development of a normal immune system. From infancy to puberty, the thymus grows and then begins to shrink and become dormant. The thymus cells in normal adults are replaced by fat as part of the aging process. Here is the big distinction in people with myasthenia gravis. The thymus gland cells in myasthenics are abnormal. Clusters of immune cells proliferate, resulting in an enlargement of the thymus. (This type of abnormal cell growth is usually found in spleen or lymph nodes during an active immune reaction.) Some MG patients also develop thymus tumors or thymomas. In most cases, thymomas are benign but they can become malignant.

It's evident that thymus abnormalities and myasthenia gravis are connected but the exact relationship and the triggering mechanism are not completely understood. One theory is that the thymus gland may provide the "wrong" instructions to developing immune cells which stimulate an autoimmune response. As a result, the production of AChR antibodies sets up the perfect environment for myasthenia gravis to develop.

Then there is the question of why an immune system goes haywire and produces such harmful antibodies. Before birth, the fetal immune system screens out certain B cells that produce antibodies that could be destructive. Perhaps some of these B cells escape the screening process in myasthenics and activate later to produce AChR antibodies.

There is also the genetic component. Are there genes that make some people more susceptible than others to develop an autoimmune disorder? And do infectious agents help to trigger a person's genetic vulnerability to a disease? For example, some scientists believe that viruses may activate the immune system to produce harmful antibodies in a case of "mistaken identity." Maybe the answer lies in some combination of these possibilities.

CATEGORIES OF MG

I soon learned that there are many types of myasthenia gravis. In fact, each case of mysasthenia gravis is unique just as each snowflake varies one from another. If you have MG, you may have one or two or more of the symptoms. You may find that your difficulties wax and wane and make it hard for you to pinpoint at times. In Part I, I described the ways in which I've been affected by MG. Now let's look at how the medical profession classifies myasthenia gravis.

Basically, myasthenia gravis is divided into three groups: ocular, generalized, and severe generalized MG. Each group has distinct symptoms but the common thread is muscle weakness in the voluntary muscles of the body.

The first group of *ocular myasthenia* generally only involves the eye muscles. People with this mild form of MG often have a sleepy expression with drooping eyelids (ptosis) and sometimes double vision (diplopia). The muscles that control eye movements may also be affected. Again, the symptoms vary from person to person with muscle weakness more evident later in the day. Ocular myasthenia is considered a mild form of the disease. While it may be a chronic condition warranting lifelong medical care, it is not fatal. Because the symptoms are confined to areas of the eye, this type of myasthenia is a localized form of the disease.

The second group of MG is called *generalized myasthenia*. And, as you can imagine from its label, this category runs the gamut and more areas of the body are involved. Some people in this category have mild symptoms with slow or gradual onset, while others experience acute and severe symptoms with rapid progression. In the milder forms of generalized myasthenia, problems may begin with weakened eye muscles and then spread to the muscles of the arms and legs. In time, the muscles for talking, swallowing, and chewing may become involved, resulting in slurred speech with a nasal tone quality (dysarthria) and difficulty swallowing (dysphagia) and chewing food. Respiratory muscles are not affected. Myasthenics with this mild form usually respond to drug therapies.

As its name suggests, *severe generalized MG* is the most serious form of the disease. In addition to the weaknesses described above, this type of MG involves the respiratory muscles. Breathing problems can generate crisis situations for the myasthenic. In addition, enlarged thymus glands are a common occurrence in patients with severe, generalized MG. (Remember that the thymus gland produces the antibodies that destroy receptor sites in the neuromuscular junctions.) And, finally, the response to drug intervention and prognosis is not as favorable as in the two previous groups. Severe, generalized MG is the diagnosis I've been living with for almost 30 years.

PROGRESSION & INCIDENCE OF MG

There is no known cure for myasthenia gravis and the course or progression of the disease is unpredictable. It may advance slowly or rapidly, both in extent and severity; it may reach a certain level and plateau for years.

In some cases, there are periods when symptoms subside. Spontaneous, partial or complete remission occurs in approximately 25% of patients, usually in the first few years. If a patient goes into remission, he or she may remain free of symptoms for years and then suddenly relapse. Sometimes this relapse includes an exacerbation of the disease.

It is impossible to say how many cases of myasthenia gravis there are in the United States due to the great number of undiagnosed, improperly diagnosed, and unreported cases. The latest and best information from the national headquarters of the Myasthenia Gravis Foundation of America estimates the number of afflicted at 70,000 or about 14 persons for every 100,000 people in our country.

However, there is some dispute as to the actual number. Most researchers and scientists in this field believe the number of people with MG is much higher. (At the time of my diagnosis, I wondered why my internist had not been able to identify my varied and puzzling symptoms. Now I understand that many family doctors never or rarely see a patient with MG.)

MG is found in all ethnic groups and in both males and females. There is a wide range of ages when MG symptoms first occur. The disease often strikes young women in their 20s and 30s, while men generally are affected in their 50s through their 80s. MG can also happen at birth or shortly thereafter, in early childhood, adolescence, early or late adulthood, and even into the ninth decade of life.

Newborn infants of myasthenic mothers may contract a form of MG which is, fortunately, temporary. The mother with MG passes antibodies to the fetus. The newborn may show signs of myasthenia but these symptoms usually subside in two or three months. Although more than one family member may have MG, the disorder is not inherited. MG is also not contagious.

When MG was first studied, it seemed that more women than men were affected by the disorder in a ratio of about 3:2. More recent studies in our country indicate that as people live longer, more men than women are being diagnosed with MG. If this trend continues, doctors expect to see more patients with MG in the future.

For patients whose myasthenia begins early in life, the disease usually shows its severity in the first three years. About one-third of these patients experience modest weakness and then gradually improve. Another one-third get moderately weak and more or less stay at that level. The remaining one-third continue on a downward spiral and get persistently weaker. In patients who develop MG after the age of fifty, the extent of the disease is usually determined in the first year. Of course, these predictions are only generalizations.

All patients notice that fever or infections make their illness worse for awhile. The disease may intensify just before and during the first part of a woman's menstrual cycle. Then too, emotional upsets can make the disease worsen temporarily. Generally, symptoms can almost always be improved with rest, at least until they become severe.

EARLY DISCOVERY OF MG

When did the medical community first learn about myasthenia gravis? If we look back in history, we find the first inklings of MG-like symptoms recorded by an English physician, Thomas Willis, in 1672. He described patients as:

"...those labouring with a want of Spirits, who will exercise local motions, as well as they can, in the morning are able to walk firmly to fling about their Arms hither and thither, or to take up any heavy thing; before noon the flock of the Spirits being spent, which had flowed into the Muscles, they are scarce able to move Hand or Foot."

Of course, Dr. Willis did not know that he was recording details of people with myasthenia gravis at that time.

Almost 30 years earlier, an American Indian chief with the distinctive name of Opechankanough exhibited myasthenia-like symptoms. The once strong and vigorous Indian chief experienced such weakness of body (including drooping eyelids) that he needed to be carried into battle by his warriors.

However, it was not until the end of the 19th century and discoveries in Germany and Poland that a giant step was taken. In 1879, the German physician Dr. Wilhelm Erb detailed patients with muscle weakness in the eyelids, facial, and neck muscles with difficulties chewing and swallowing. Fourteen years later (1893) in Poland, Dr. Samuel Goldflam, citing the work of Erb and his own findings, summarized the characteristics of a muscle weakness disorder which was tentatively named Erb-Goldflam Symptom-Complex.

Dr. Goldflam's description was considered to be "the most important ever written in the history of the disease," according to Dr. Henry Viets, the founder of the first clinic for MG in the United States.

A year or so after Dr. Goldflam's findings, Dr. Friedrich Jolly renamed the disease *myasthenia gravis pseudoparalytica*. Dr. Jolly also made the incisive suggestion that myasthenia involved a problem in neuromuscular transmission—a remarkably astute finding. Dr. Jolly also proposed that this disorder of muscle weakness might be helped by a drug called physostigmine salicylate, an antidote for paralysis brought on by curare poisoning. Oddly enough, there is no record that Dr. Jolly treated any patients with this antidote.

(In the next chapter, we will discuss myasthenia gravis treatments and therapies and discover when this drug was first used.)

In 1901, Dr. William Osler of Johns Hopkins University in Baltimore, Maryland condensed the name of the disease to the more manageable *myasthenia gravis*. So, by the early 1900s, MG was known to exist and the search for a greater understanding of the disease was underway. Pathologists doing anatomical studies of deceased MG patients discovered an interesting finding. In some of the autopsies, the thymus gland was enlarged and often there were tumors present in the gland. Yet this was not the case in *all* MG patients. Still the link between the thymus gland and myasthenia was established. So even though myasthenia gravis was first described in the 1600s, its symptoms were only clinically designated as MG in 1895.

In those early days, MG was sometimes fatal. Today, there are a number of different treatments available. But, of course, it cannot be treated if it isn't diagnosed.

The unfortunate situation for perhaps thousands of people afflicted with myasthenia gravis is that the disease is misdiagnosed for long periods of time.

In the next chapter, we'll examine how MG is diagnosed and continue the story of the treatments and therapies currently available. Along the way, we'll meet some medical detectives and extraordinary individuals.

FACTS IN A NUTSHELL

In this chapter:

- MG is defined as a chronic, autoimmune neuromuscular disease with varying degrees of muscle weakness. The crucial "glitch" occurs in the neuromuscular junction (gap) between nerves and muscles.
- Antibodies block or destroy ACh receptor sites on muscle tissue in myasthenia gravis, thus preventing or obstructing the normal transmission of acetylcholine for muscle contractions.
- What causes the immune system to produce harmful antibodies? What is the connection between the thymus gland and MG? How is genetic vulnerability involved in myasthenia gravis? These are some of the questions yet to be answered.
- MG can be categorized as ocular, generalized, and severe generalized. Each type of myasthenia gravis has its distinct symptoms, although each case and its progression are unique. The incidence of 70,000 myasthenics is considered grossly underestimated due to the prevalence of undiagnosed and misdiagnosed cases.
- MG was described in the 17th century and, treatments began in the late 1800s. In 1901, the classic symptoms of the disease were identified as myasthenia gravis.

181

CHAPTER 10

UNRAVELING TESTS, TREATMENTS & THERAPIES

While we've been on a journey of my experiences with myasthenia gravis, many medical terms have been mentioned. Perhaps you are at the beginning of your own journey or you are seeking greater understanding of what is involved in MG care. This section deals with the tests used in diagnosing myasthenia gravis and the treatments and therapies currently available. It is more complete than the course of medical tests and treatments I have experienced and recounted in the first part of this book.

You'll find out why certain procedures are done as well as how specific medications and therapies work. I hope you'll find it a helpful reference. And, as always, I'll try to make it clear and understandable.

Please be aware that this section is for your information. Only your physician and other medical professionals can advise you as to your unique needs. I am not endorsing or recommending any medical test, treatment or therapy.

TESTING FOR MG

A comprehensive physical examination which includes a complete medical history needs to be taken before undergoing tests for any illness. Your doctor may then refer you to a specialist.

When there is the possibility of myasthenia gravis, a consultation with a neurologist is usually recommended. As a result of this examination, the doctor may then indicate the need for additional testing to confirm the diagnosis.

Once the neurologist examined my symptoms and listened to my description of ongoing muscular weakness, he then proceeded to test me for myasthenia gravis by means of the Tensilon test and an EMG or electromyography. Another test used in confirming MG is a blood test which measures the levels of ACh receptor antibodies.

You or your family member or friend may have experienced one or more of these procedures in diagnosing MG. Or perhaps you're preparing to take one of these tests after undergoing a thorough neurological examination. It's helpful to understand what is happening during each of these tests, how they work, and what the results indicate. Here's what I have learned about these procedures that give physicians the information necessary to identify myasthenia gravis.

183

The Tensilon (Edrophonium Chloride) Test

In my own case, an injection of Tensilon (edrophonium chloride) produced a very dramatic change. My slurred speech, drooping eyelids, and extremely weak muscle tone in my body improved a hundredfold—as if by magic.

What causes this amazing turnaround? The Tensilon blocks the action of acetylcholinesterase, the enzyme that breaks down acetylcholine. Since more acetylcholine (ACh) is available to transmit nerve impulses to muscles, the action of ACh is given a boost. Weak muscles become stronger for several minutes.

This benefit is worthwhile as a testing agent but not as a lasting treatment. In fact, I recall that I was encouraged to have a nourishing meal during my brief reprieve. (I had not been able to chew or swallow solid foods.) The doctors, nurses, and other health professionals who witnessed my "return to normal" were interested in seeing a positive response to the Tensilon test. Some of them had not seen the striking difference that one simple injection can make.

In testing infants and children, physicians sometimes use a drug called neostigmine because it works for a longer period of time. The action of Tensilon can be too brief for an accurate test in small children. Because of its simplicity, the use of Tensilon or neostigmine is often the first diagnostic tool used once a thorough neuromuscular exam points to MG.

Testing for AChR Levels

A blood test can confirm the diagnosis of MG by measuring the level of AChR, antibodies that destroy or block acetylcholine receptor sites. In myasthenics, the immune system mistakenly produces antibodies that block the transmission of acetylcholine from nerve impulses to muscles. It would seem logical that all myasthenics would have high levels of antibodies blocking AChR. In reality, the picture is somewhat blurred. About 80% of myasthenics with generalized MG and 54% of patients with ocular myasthenia have appreciable levels of AChR antibodies.

There is a wide variation of these antibody levels in patients with MG. The amount of AChR antibodies in the blood may be minimal when a person first shows signs of myasthenia and then become elevated as the disorder progresses. Then there are some myasthenics (about 10%) who do not test positive for AChR antibodies but have other antibodies that affect the acetylcholine receptor sites. It is important to note that a "normal" or negative antibody level does not necessarily mean that a person is free of myasthenia.

To add to the confusion, significant levels of AChR binding antibodies are also found in patients with systemic lupus erythematosus, amyotrophic lateral sclerosis (ALS), and thymomas without myasthenia gravis. There have even been positive antibody levels found in patients taking muscle relaxants or after general anesthesia. Again, this blood test for AChR antibodies is a useful and accurate tool in the medical kit. But it must be combined with other diagnostic tests to confirm the diagnosis of MG.

185

Electromyography

You may recall that I found the electromyography (EMG) test particularly difficult. But that was many years ago. Laboratory techniques have progressed. Basically, this test involves the use of special equipment, an electromyogram, to stimulate muscle tissue by means of electrical impulses. By observing the patient's muscle responses to nerve stimulation, the technician can determine whether the muscle is responding within the normal range or not.

There are two kinds of EMG testing: repetitive and single fiber nerve stimulation. In repetitive nerve stimulation of most myasthenics, the muscle response weakens as the stimulation is repeated. While this helps to identify MG patients, positive repetitive nerve stimulation is found in other neuromuscular diseases. And, to add to the variability, people with a mild or ocular myasthenia may have a normal response to repetitive nerve stimulation.

The second test of single fiber EMG is the more sensitive procedure. Almost all myasthenics show impaired neuromuscular transmission when tested by this method. It's also useful because it helps to exclude those individuals who do not have myasthenia gravis. But, once again, a positive test may indicate MG or some other neuromuscular disorder.

In addition to Tensilon, AChR antibody blood levels, and EMG testing, your physician may also order other tests. Specialized equipment such as a CT scan (computed tomography) or an MRI (magnetic resonance imaging) may be used to identify an abnormal thymus gland or thymomas (tumors) on the thymus gland. A pulmonary function test is also useful to measure the strength of the breathing muscles.

While the different testing procedures are helpful in making a diagnosis, all means of discovery begin with a review of the patient's medical history and physical and neurological examinations. Once the diagnosis of MG is confirmed, the next step is finding the right treatment for the patient. In my own case, I found that much patience was needed before the right combination of treatment and therapy was discovered. At the same time, I wanted to know as much about the medications prescribed and their potential side effects.

MEDICAL TREATMENTS FOR MG

Although MG is not curable, it can be controlled. The drugs used to treat MG were developed as a result of learning the mechanism of myasthenia. Remember that myasthenics produce antibodies which destroy muscle receptor sites needed in the transmission of chemical messages from nerves to muscle cells. Many of the treatments for MG are geared to inhibit the production of those antibodies. In order to accomplish this suppression, treatment may require a combination of medications and surgery.

Treatment of myasthenia gravis focuses on several different courses. All are effective in alleviating the symptoms of MG. They differ, however, in the side effects they cause. Let's look at the MG medications with the idea of understanding how they work to relieve muscle weakness.

Anticholinesterase Drugs:

Mestinon, Prostigmin, Mytelase

As its name suggests, the medications in this category block the breakdown of acetylcholine. More acetylcholine (ACh) provides more of the chemical messenger needed for muscle contraction, a benefit that increases muscle strength.

In 1895, Dr. Friedrich Jolly described myasthenia gravis as a problem with neuromuscular transmission and suggested that a certain drug, physostigmine salicylate, might be a helpful in treating MG. But as we mentioned, there is no record that the good Dr. Jolly ever tested his theory.

It was not until the 1930s that a young Scottish physician, Dr. Mary Walker, working in a hospital in Greenwich, England, conducted the experimental use of physostigmine as a treatment for an MG patient. Dr. Walker's experiment was based on the fact that MG symptoms were similar to curare poisoning. Since physostigmine was the known antidote for curare poisoning, she reasoned that it might also benefit myasthenics. Dr. Walker's experiment with treating a patient with physostigmine proved successful in temporarily relieving muscle weakness. Soon after her groundbreaking discovery, the drug Prostigmin was synthesized. It proved to be an effective and safer drug than physostigmine, and offered MG patients a much-needed medication. (Dr. Walker's patient and Jean Welch Kempton were pioneers in the treatment of MG. Ms. Kempton described it well in her book, *Living with Myasthenia Gravis.*)

Then, in the 1950s, edrophonium chloride, Tensilon, was discovered to be a tool for diagnosing MG. Remember that the Tensilon test blocks the enzyme acetylcholinesterase which breaks down acetylcholine (ACh). More ACh translates into more and greater control of muscle contraction. But the Tensilon injection has the drawback of being very transitory. For this reason, it is used as a diagnostic tool rather than a long-term treatment.

Another landmark medical discovery occurred in 1954 when the drugs Mestinon and Mytelase were introduced. The drugs Mestinon, Prostigmin and Mytelase are in the same anticholinesterase category as Tensilon but they have the added benefit of working for longer periods of time. These drugs are usually the first method of treatment for myasthenia gravis.

Generally, this type of drug increases muscle strength one-half hour after taking the medication and loses effectiveness after three or four hours. There are also timespan drugs in this category which release the medication over a longer period of time. This is particularly useful for those who wake up with severe weakness. By ingesting a Timespan anticholinesterase tablet at bedtime, the patient can feel a "muscle boost" by morning. This type of extended release drug is not given during the day time because of the waxing and waning of muscle strength.

Finding the most beneficial dosage for an individual patient is a challenge for the physician. The dose must be carefully regulated and the myasthenic's cooperation and patience is needed. An underdose will be ineffective; an overdose could result in severe weakness and even death. Shortness of breath or difficulty swallowing shortly after taking one of these drugs is a sign of an overdose. Immediate medical care is needed.

Unfortunately, as helpful and effective as these medications are, they can also be quite toxic and produce many side effects. The most common adverse reactions are gastrointestinal problems. An overactive digestive tract may cause cramping and diarrhea. The patient may also experience an increase in perspiration, muscle twitching, heart palpitations, and urinary frequency.

Another drug that is sometimes used in treating MG is ephedrine sulfate. In the late 1920s, a young woman, Harriet Edgeworth, studying at Rush Medical School was diagnosed with myasthenia gravis. Ms. Edgeworth had a doctorate in chemistry and thought that ephedrine (which is chemically similar to adrenaline) might be helpful in treating her MG. Unfortunately, she wasn't able to convince her physicians. But after receiving her medical degree, she accidentally discovered the benefits of ephedrine when she was taking ephedrine with another drug for menstrual cramps. She published her findings in 1930 and subsequently used ephedrine for many years to treat her myasthenia. While ephedrine is still used today, often in combination with other drugs, it does have its downside. Some of the common side effects include insomnia, nervousness, and heart palpitations.

Immunosuppressant Drugs

If the immune system is overactive and producing antibodies that block muscle contraction or any other bodily function, a logical way of combating the situation is to slow down the immune system. Immunosuppressant drugs act as the name implies by suppressing or inhibiting the immune system.

You are probably most familiar with the steroid immunosuppressant medication called prednisone. Chemically, prednisone is the synthetic version of the hormone cortisone which our bodies produce.

In the 1960s, synthetic hormones, steroid drugs, were developed. Steroid drugs have a two-fold effect: they decrease inflammation and suppress the immune system. By limiting the production of abnormal antibodies in myasthenics, more ACh transmission between nerve and muscle cells is possible.

The drug prednisone is the most commonly used immunosuppressant medicine in treating myasthenia gravis, especially those patients who have severe, generalized MG. The good news is that 90% of MG patients using prednisone experience a positive and definite improvement.

However, the benefits of prednisone are offset by many side effects, including an increased appetite and rapid weight gain. In some patients, blood sugar levels may become elevated and trigger an underlying mild diabetes. Blood pressure may increase and the heart may be under added strain because of fluid retention in the body. Steroid users over the age of fifty often suffer cataracts. Thinning of the bones (osteoporosis) can also occur because steroids block the uptake of calcium from the diet. In addition, there is an increased risk for infection. And, finally, steroids can worsen emotional problems as these drugs sometimes promote feelings of depression.

Despite all the negative side effects, steroid drugs are still considered to be "wonder drugs" in many cases. But they need to be used with careful physician monitoring. Your doctor will also guide you in tapering off the dosage of a steroid drug such as prednisone. This kind of medication should never be stopped "cold turkey."

In some instances, prednisone is ineffective or needs to be discontinued for some time due to adverse side effects. Other immunosuppressant drugs which are non-steroids may be prescribed. Imuran, Cytoxan, and Sandimmune are strong medications capable of controlling MG by blocking the production of AChR antibodies.

Imuran (the brand name for azathioprine) is a more potent and expensive drug than prednisone. Improvement may be very slow. Several months may pass before substantial benefit can be measured. During this time, the patient needs to have red and white blood cell counts as well as liver enzyme levels monitored. Although Imuran does not have the diversity of side effects found in taking prednisone, it does have serious drawbacks. Reactions such as fever, abdominal pain, and nausea should be reported to your physician and may require discontinuing the drug.

Cytoxan (brand name for cyclophosphamide) is a stronger immunosuppressant drug than Imuran. In fact, Cytoxan is such a strong immunosuppressant that it is used as a chemotherapy agent in cancer treatment. In myasthenics, Cytoxan is prescribed when other drugs or therapies have failed to produce results. Once again, the side effects are numerous.

Many patients who are taking the drug Cytoxan experience thinning of the hair, nausea, decreased red and white blood cell levels, and an increased risk for all kinds of infections—including flu and shingles. While the side effects are severe, the long-term benefit may be substantial.

In a number of patients taking Cytoxan, the disease improves markedly within two to six months. Long-term, severe myasthenics may require several months to a year before benefit is apparent. Although Cytoxan is sometimes effective in controlling the illness, the goal is always to seek the lowest dose or to substitute a less risky medicine once control has been obtained.

Another immunosuppressant drug called Sandimmune (the brand name for cyclosporine) is used for organ transplant patients. It is currently being studied as a possible drug for myasthenia. Some of the adverse effects of Sandimmune include high blood pressure, headaches, kidney problems, and increased body hair.

Intravenous immune globulin (IVIg), which is also known as human gamma globulin, is another type of immunosuppressant drug used to treat myasthenia patients. IVIg is produced from the screened blood of multiple donors. Intravenous injections of IVIg are especially helpful for individuals in an acute stage of myasthenia. It is usually administered in a series of infusions given for several days.

Exactly how IVIg works is unknown, but it is believed to inhibit the antibodies that destroy ACh receptor sites and provide a flood of "beneficial" antibodies. Unfortunately, while the success rate of IVIg is impressive, the positive effects only last for weeks or a few months at most.

The side effects of IVIg include headache, skin rashes, and flu-like symptoms. Another concern is the potential formation of blood clots due to the high concentration of protein in gamma globulin. In addition, IVIg is a very expensive medication.

While the above medicines are prescribed in the treatment of MG, none of them is a cure-all. Many of them have serious and long-term side effects. Often the medications for myasthenia need to be supplemented with various therapies including a plasma exchange process that eliminates harmful antibodies and surgery to remove the thymus gland. Let us explore some of these procedures.

MG THERAPIES

Plasmapheresis: (Plasma Exchange)
Another way to attack the malfunction of the immune system is a procedure of "drawing off plasma" called plasmapheresis. This technique involves removing blood from the patient and then separating the blood cells from the plasma or liquid portion of the blood. After AChR antibodies are withdrawn from the blood

plasma, the blood cells and "cleansed" (or artificial) plasma are transfused back into the patient. Generally the artificial or substitute plasma is a saline solution with albumin protein.

This specialized plasma exchange usually needs to be repeated several times with each session lasting a few hours. After several treatments, the patient's symptoms are often remarkably better, sometimes eliminating all evidence of illness for several months. The benefit, however, is temporary.

While plasmapheresis may be useful, it must be combined with drugs that control the disease. If not, patients revert to their previous level of symptoms. There is also the potential for serious allergic reactions and headaches to occur. Careful monitoring and plenty of fluids help to minimize these side effects. In addition, plasmapheresis is very expensive. It is usually reserved for those who have not responded to other therapies or those in acute or crisis situations. In some instances, plasmapheresis is used to stabilize patients before surgery to remove the thymus gland.

Surgery—Thymectomy

Because of the severity of my condition when I was diagnosed, my neurologist advised an immediate thymectomy, surgery to remove the thymus gland. This is major surgery. In my own case, an incision was made lengthwise on my chest and the breastbone (sternum) was cut, as in open heart surgery. Since the chest cavity was exposed, the surgeon was able to remove the entire thymus gland.

For some time, scientists have tried to discover a less invasive thymectomy surgery. The latest attempt involves a technique of robotic-assisted thymectomy which I mentioned in the Postscript to Chapter 2. Much of the present research compares the effectiveness of the two surgical techniques. Whichever method is used, the important element must be the complete removal of the thymus and surrounding tissue to avoid having to repeat the surgery.

Why remove the thymus gland? Around the turn of the 20th century, physicians doing postmortems on myasthenia patients discovered an interesting phenomenon. Many of these patients had enlarged thymus glands or tumors (thymomas) on their thymus glands. Could the thymus gland be the "culprit" in MG?

Early scientists hypothesized that the thymus gland played an important role in myasthenia and that excising the thymus gland might be helpful in easing the symptoms of MG. The first successful removal of a thymus was performed in 1911 by the Swiss surgeon, Dr. Ernst Ferdinand Sauerbruch. His patient had an enlarged thymus. After surgical removal of the thymus gland, the patient showed remarkable improvement.

Later surgeries targeted myasthenics with tumors on their thymus glands (thymomas). In 1939, an American surgeon, Dr. Alfred Blalock, was the first physician to remove the thymus gland of a patient with thymomas. Remarkably, the young woman myasthenic had a complete remission within three years. Several years later, Dr. Blalock performed thymectomies on MG patients who did not have enlarged thymus glands or thymic tumors. The majority of his patients showed improvement.

Given the history and the positive results of thymectomy, you may be wondering why this surgery is not the "magic" answer. Even though there have been a number of success stories, thymectomy is not a universal remedy. In fact, there has been an ongoing controversy over its effectiveness. According to a statement issued by the Myasthenia Gravis Foundation of America:

"The response to thymectomy is variable; about 30% eventually experience complete drug-free remission, and another 50% experience significant improvement."

At the time of the discovery of abnormal thymus glands in MG patients, the function of the thymus gland was poorly understood. Today, we know that the thymus gland produces antibodies as part of our immune system during early childhood. The thymus is a butterfly-shaped gland located in the chest behind the upper end of the breastbone, along the trachea and just below the thyroid gland. It develops rapidly in the infant, grows more slowly around puberty, and then gradually withers away in adulthood.

In most myasthenics, the thymus is enlarged, stimulating antibodies that destroy ACh receptor sites. In some patients with MG, the thymus gland may also have tumors (thymomas) which are usually benign. By removing an enlarged or tumorous thymus gland, the production of harmful antibodies can be reduced. For this reason, the surgical removal of the thymus gland, a thymectomy, can be a very effective therapy. Some of the benefits of a thymectomy are a decreased need for medication and an increased chance of remission. It may also lessen the incidence of cancer.

(Currently there is an international study of the pros and cons of thymectomy in patients with or without thymomas. This is discussed in Chapter 12.)

You may wonder what initiates or promotes the destructive activity of producing harmful antibodies in an enlarged thymus gland. Some possible "triggers" are infection, inflammation or an allergic reaction within the thymus gland. Removing the thymus gland offers a one time surgical risk in hopes of improving the patient's symptoms and decreasing the need for medication.

While we have surveyed the current tests, treatments and therapies available, much work is being conducted in laboratories and clinics all over the world. In the next chapter, you will find many resources to contact for more information and the latest updates on treatment.

FACTS IN A NUTSHELL

- The principal tests for diagnosing MG are: the Tensilon test, electromyography (EMG) and the measurement of AChR antibodies levels in the blood.
- Medications for MG are classified according to the way in which they relieve muscle weakness. Anticholinesterase drugs block the breakdown of acetylcholine, thus providing more of the chemical messenger needed for

muscle contraction. Immunosuppresant drugs repress the immune system, inhibiting the production of harmful antibodies.

- Therapeutic procedures used to treat MG include plasmapharesis, the cleansing of harmful antibodies from the liquid portion of the blood, and thymectomy, the surgical removal of the thymus gland.

CHAPTER 11

SEEKING AND USING RESOURCES

A pebble dropped in a lake of still water sends out ripples that reach far from the point of entry. In much the same way, dealing with a chronic illness is not a solitary challenge. There are "ripples" of people and "currents" of organizations and "tides" of information that can change the flow of our journey. Our lives are not in a vacuum; we are not alone. As the poet John Donne wrote so well, "No man is an island, entire of himself..."

Yet illness can cause us to feel isolated. Even the very word *disease* connotes a person "not at ease" with one's self. Still there are numerous resources that can enrich our lives if we seek them out and take advantage of them. Here are some suggestions for learning more about MG or other chronic illnesses, and tapping the abundant and myriad resources in your community, family, and friends. Ultimately though, *you* are your greatest asset. You hold the key to responding well to the challenge of a chronic illness.

CHOOSING A DOCTOR

In the beginning of this book, I described my difficulty finding a physician to diagnose my condition. For a number of years, I relied on a highly recommended doctor's guidance and lost precious time. Later on I discovered that many physicians (non-neurologists) never or rarely treat a patient with myasthenia gravis.

Because of the nature of the disorder and the ongoing need for care, your choice of medical assistance is crucial. While you may begin with your family practice or internal medicine doctor, you should seek the help of a neurologist who has experience treating myasthenia gravis. I have great respect and admiration for medical professionals. It takes a dedicated person to undergo the grind of medical school, residency, and fellowship before passing the Medical Boards of his or her specialty.

Your choice of a physician may be limited by your health insurance plan or HMO. Check with your plan for a list of neurologists in your area. You can also ask for referrals from your primary physician, other professionals in the healthcare field, family, and friends. Other search possibilities are your hospital's physician referral service and a medical association's list of members.

Once you have a list of neurologists in your area, you can then do some research on individual physicians. For example, you can learn if a doctor is board certified by contacting the American Board of Medical Specialists. A board certified physician has specialized training and has passed a thorough exam evaluating his or her knowledge in a particular field. While there are many fine physicians who are not board certified, this credential adds a certain level of confidence in the doctor's dedication and ability to provide competent care.

The web site for the American Board of Medical Specialists is:

www.abms.org/newsearch.asp

To find a neurologist in your area or discover more about a particular neurologist, you can search the American Academy of Neurology web site at:

www.aan.com

Once you have completed your information gathering, you are now ready to select a physician. Be sure that the doctor you've chosen works with your insurance plan and is also taking on new patients. If possible, schedule an initial meeting to discuss your symptoms, concerns, and needs. Be attentive to the interaction between yourself and the physician. Does the doctor take the time to listen and explain? Of course, a mutual respect and bond develops over time. But you can often sense whether or not this professional relationship will "work" or not.

It's important to keep in mind that *you* are a vital part of a good doctor/patient relationship. You need to supply the physician with the information he or she needs to effectively treat you. This communication entails describing your symptoms, providing a current health record including medications, dosages, and allergies, and being honest about alternative medicines or other areas in your life which may be affecting your health. Make a list of questions that you need answered and give the physician the results of any tests pertinent to your illness. In other words, provide the doctor with the tools necessary to provide quality care.

If you think of your interaction with you doctor as a partnership, you're on the right track. And, of course, this also means that you follow the physician's directions and advice and keep him or her informed of any changes in your condition. Taking responsibility for your health by means of a trusting partnership with your doctor will help you physically and emotionally.

BROWSING THE WEB

We live in an online world of information. Recently, a friend of mine told me that her father had strange symptoms that seemed to defy diagnosis. Her first instinct was to do an internet search matching her Dad's symptoms to known illnesses. Within a short time, she stumbled on the words *myasthenia gravis*. After bringing this information to her father's physician, her Dad was properly diagnosed and is being treated for MG.

While the internet is a valuable tool and resource, self-diagnosis is not recommended. In fact, great care needs to be taken in using the internet. A superb source of using the internet resources wisely is *The Official Patient's Sourcebook on Myasthenia Gravis: A Revised and Updated Directory for the Internet Age*, James N. Parker, M.D. and Philip M. Parker, Ph.D., Editors, Icon Health Publications, San Diego, CA, 2002. It's an extensive guide to researching MG books, periodicals medications, nutrition, and clinical trials, as well as finding medical libraries and knowing your rights and responsibilities as a patient. I think you'll find it an invaluable resource. Icon Health Publications also publishes sourcebooks for a number of chronic illnesses. Check your online bookseller or contact the publisher directly at:

www.icongrouponline.com/health

While you could search *myasthenia gravis* on any of the internet's search engines and find information, it is vital to find reliable sources. Some web sites recommended in *The Official Patient's Sourcebook* are:

The National Institutes of Health (NIH):
www.nih.gov/health/consumer/conkey.htm

The National Library of Medicine (NLM):
www.nlm.nih.gov/medlineplus/healthtopics.html

The National Institute of Neurological Disorders and Stoke (NINDS):
www.ninds.nih.gov/health

Each of the above sites is part of the National Institutes of Health, the leading Federal research center in the United States.

You'll also find specific and clear information on myasthenia gravis at the following web site for the **Myasthenia Gravis Foundation of America** at:

www.myasthenia.org

The Myasthenia Gravis Foundation is an organization totally dedicated to providing practical information and supporting research with the goal of conquering MG. I'll describe more of their services in the next two sections.

FINDING BOOKS AND ARTICLES

When I was first diagnosed with MG, I immediately sought first-person account books. With the rare exception of Jean Welch Kempton's book, *"Living with Myasthenia Gravis: A Bright New Tomorrow,"* I was sorely disappointed. One of the reasons I continued to work on this book for so many years is the need to share my lifetime of experience with MG. I'm not looking for any personal glory. My mission is much simpler. I believe there is a hunger for the written word of "one who has been there," especially since I've "been there" for a protracted length of time.

Perhaps you have been recently diagnosed with a chronic illness and are struggling with the challenge. Or maybe you have had MG for a number of years but you're looking for a fresh approach, new ideas to stimulate your own path. I hope my book fulfills some of the needs of those with myasthenia gravis or other chronic illness. I feel compelled to reach out and try to ease your journey.

For books on myasthenia gravis, you might want to start with online book sources such as the web sites for Barnes & Noble at: **www.bn.com** and Amazon at **www.amazon.com**. You'll find a listing of available books as well as reviews from critics and other readers. Most of the personal accounts of MG are out-of-print or very dated.

One excellent book by an MG patient and physician is *Attacking Myasthenia Gravis: A Key in the Battle Against Autoimmune Diseases* by Ronald E. Henderson, M.D., Court Street Press, Montgomery, AL, 2003. Dr. Henderson offers a well-written account of his experience from the perspective of a physician and entrepreneur. He has sounded a call for the medical profession and communities to become involved in establishing programs to meet the challenge of the chronically ill. Dr. Henderson has also initiated a foundation for researching the causes and treatment for myasthenia gravis and other autoimmune diseases.

Your local library is an excellent source for obtaining books and magazine articles on myasthenia gravis. Interlibrary loans extend from your county of residence to regional searches. You may find it beneficial to solicit the aid of the reference librarian to access the computerized catalog system effectively. In most instances, you can also connect to your library's resources by means of your home computer.

For recent news releases, you can browse the **MEDLINE/plus** web site. Another great source to help you keep current on new trends is the clear, concise, and free newsletter produced by the Myasthenia Gravis Foundation of America, Inc.

JOINING SUPPORT GROUPS

As soon as I was well enough, I joined a local chapter of the Myasthenia Gravis Foundation of America (MGFA). This organization was founded more than fifty years ago by Jane Ellsworth, the mother of a teenager with myasthenia gravis. When her daughter was diagnosed with MG, Ms. Ellsworth felt stymied by the lack of information about myasthenia. Encouraged by her daughter Pat, Ms. Ellsworth worked on establishing a support group that evolved into a nonprofit, voluntary health agency. Today the MGFA, based in St. Paul, MN, provides patient services for over 100 support groups, funds research, and offers informational literature as well as an active web site. The overall aim is to help patients understand and manage MG.

Over the years, I have served on the Board of the Myasthenia Gravis Foundation of Illinois (MGFI)—a chapter of the national organization—in various capacities and am now the Secretary. Since most of the Board members are myasthenics, you can imagine the effort each exerts to participate.

As a local chapter, patient service is our top priority. We also support MGFA as well as research projects, especially those in the Chicago area. Our brochure describes the Foundation's goals:

- To support MG patients and their families with information and understanding so they can achieve productive and rewarding lives.
- To provide access to a prescription program that offers reduced rates on costly medication.
- To inform the public about MG and promote greater awareness and understanding.
- To provide physicians information on the latest developments in MG research in order to avoid misdiagnosis and delay in treatment.
- To work with nurses and help them understand and respond to the special needs of MG patients.
- To support and assist in research to find the cause and discover the cure for MG.

Of course, some of us are "joiners" while others tend to be more comfortable on the sidelines. It does take some courage to face the reality of having MG and needing support from others. Once again, you are not alone in your concerns. Many of the members battled similar feelings of hesitancy. In some instances, they were worried that a support group might even be depressing. Fortunately, most members discover that these fears are unfounded.

While a support group may not be for everyone, it does have the advantage of surrounding you with others traveling the same road. It's amazing what we can learn from one another.

The MGFA provides a fountain of information and an ongoing educational forum in 36 chapters and 100 support groups. Its March 2007 mission statement describes it well:

"MGFA is committed to finding a cure for myasthenia gravis and closely related disorders, improving treatment options and providing information and support to people with myasthenia gravis through research, education, community programs, and advocacy."

I heartily recommend that you seek out your local chapter and become involved in the group. To find a chapter in your area, contact:

The Myasthenia Gravis Foundation of America (MGFA)
1821 University Ave. W, Suite S256
St. Paul, MN 55104-2897
(800) 541-5454
(651) 917-6256
http://www.myasthenia.org

Another association dedicated to supporting research and patient advocacy is the **American Autoimmune Related Diseases Association, Inc.** This organization works to raise awareness of the more than 80 autoimmune diseases including multiple sclerosis, ALS, and myasthenia gravis. You may contact this group at the web site:

www.aarda.org

For those unable to attend meetings, there is an internet chapter of MGFA which sponsors weekly chat rooms. To access this support group, go to:

www.mgfa-mgnet.org

Another web site on the internet that provides chat rooms, newsgroups, and bulletin boards is **Myasthenia Gravis Links.** This site also provides connections to the latest research in the United States and throughout the world. You can access this comprehensive online service at:

www.pages.prodigy.net/stanley.way/myasthenia

TAPPING FAMILY AND FRIENDS

What would we do without the caregivers in our life? When we discuss the professional support available or consider the tremendous amount of information which grows each year, we cannot ignore the sustenance of those closest to us. Our family members and friends are invaluable resources in our journey to wellness.

As an independent woman who likes to be "in control" of situations, I have had to learn to accept the fact that I need some help at times. Fortunately, I have a remarkable support system of people who are more than willing to give me that extra boost when I need it. Even so, I have had to adjust my thinking to *accept* assistance. It takes a fair amount of humility to recognize need and then be willing to receive it.

Often our family and friends want to be of assistance but they're not sure what is required. You may need to guide them to learn about MG and be prepared to lend you a hand. Share some of the MG literature from the MGFA or have them borrow this book to acquaint them with myasthenia. You can also invite them to go to an MG chapter meeting with you.

Recently MGFA has partnered with an internet organization called CaringBridge. It is a free web site that helps to connect MG patients with their family and friends during treatment, recovery or health crisis. To set up your web site, go to:

www.caringbridge.org/myasthenia

With families and friends spread throughout the country, this would be a convenient way to keep in touch and let others know how you're doing.

You'll find pertinent suggestions for involving your family and friends in the MGFA booklet, *Myasthenia Gravis Survival Guide: A Guide to Patient-Directed Health Management* by Jeanne Rhynsburger, R.N. One example given in the booklet is preparing for an emergency by setting up a distress phone signal to use if you're not able to talk. You may also want to form a response system according to the severity of your condition.

This practical booklet also offers many ideas about how you can organize your life to help you deal with everyday activities from caring for yourself to shopping to conserving energy. In the final analysis, *you* are your greatest resource.

FACTS IN A NUTSHELL

This chapter focuses on:

- The significance of choosing a doctor and specific methods to succeed in building a constructive partnership with a medical professional.
- Using the internet for information with the guidance of *The Official Patient's Sourcebook* and the Myasthenia Gravis Foundation's web site.
- Ways of locating books and articles on myasthenia gravis by means of online book distributors, web sites, and local library resources.
- Description of available support groups, especially the chapters of the Myasthenia Gravis Foundation of America (MGFA), online chat rooms, newsgroups, and web site bulletin boards.
- How to keep family and friends informed about MG; providing guidance on how to give you assistance when necessary; using the CaringBridge web site to connect with others; and learning to humbly accept help from your support system.

CHAPTER 12

RESEARCHING THE PUZZLE: HOPE FOR THE FUTURE

Have you ever spent hours "solving" a 5,000 piece jigsaw puzzle? At first, the task seems daunting. Then you see that certain shades of color, slight variations in tints, and distinctive shapes match definite sections within the finished product. It can be an enjoyable task of endless hours or a never-ending source of frustration. Perseverance is needed to sort through the numerous varieties of shapes and subtle changes in color. As you discover the perfect matches, draw closer to completion and the number of puzzle pieces diminishes, there is a great feeling of satisfaction. And then there's the delightful feeling of accomplishment and relief when the goal is reached.

In the scientific arena, researchers spend countless hours and lifetimes attempting to fit the myriad pieces of data together to find solutions to the mystery of diseases. Medical research scientists are the unsung heroes in the ongoing struggle to find answers and help those who daily battle serious illnesses.

Myasthenia gravis holds a unique and fortunate position as the most well understood of all the autoimmune, neuromuscular diseases. Yet much work remains. In this chapter, you will find an overview of some of the latest research findings for myasthenia gravis, the dilemmas in need of resolution, and the prospects for the future.

THE TEAM APPROACH TO RESEARCH

I'd like to begin with a husband and wife team of research scientists that I met through the Illinois chapter of the MGFA. Matt Meriggioli, MD and Julie Rowin, MD are leading neurology physicians at the Neuromuscular Center of the University of Illinois Medical Center, and members of the Medical Advisory Board of MGFI. In addition to treating the largest number of MG patients in the state of Illinois, they are also involved in basic research and clinical trials of new drug approaches in treating MG.

Their collaboration is aided by working with a group of multidisciplinary scientists. Dr. Meriggioli's laboratory is located in the department of microbiology and immunology and close to the departments of anatomy and cell biology—all within easy access for brainstorming potential links to new discoveries. This open door policy and sharing of information are vital components of their research.

One of the problems Drs. Meriggioli and Rowin are trying to resolve is how the immune system regulates itself. A clearer understanding of this mechanism would help researchers find safer alternatives to immunosuppressants such as the steroid prednisone. While immunosuppressant drugs decrease immune system activity, they are not specific and can cause serious side effects. Some of the concerns are bone loss, weight gain, depression, and an increased risk of infection. At the same time, medical drugs categorized as immunosuppressants are effective in stifling the wayward immune response, restoring receptors and allowing muscle membrane to repair itself.

The Meriggioli-Rowin team is currently conducting trials on a new immunosuppressant drug—CellCept (generic for mycophenolate mofetil) with the hope of finding a treatment with fewer side effects and greater tolerance by MG patients. These clinical trials are part of an ongoing effort internationally and at nine other U.S hospitals. While CellCept has been approved for patients of organ transplants, more clinical trials are needed to assure its efficacy in the treatment of MG.

Some CellCept studies are experimenting with varying the dosages and timing of the drugs. In some trials, prednisone is used to stabilize the patient and then CellCept is added while the dosage of prednisone is decreased. The goals of these studies are to increase muscle strength, decrease AChR antibody levels, and improve the quality of life in MG patients.

(You will find more information on CellCept research on p. 218 in the section: The Immune System—A Closer Look.)

THYMUS SURGERY OR NOT

Some myasthenics have a tumor or a growth (thymoma) on their thymus glands. Patients who have thymomas—even those without myasthenia gravis—usually undergo a thymectomy, the surgical removal of the thymus gland. This is necessary to prevent difficulties in heart function and breathing.

For the past 60 years, the thymectomy procedure has been a standard practice for MG patients with or without thymomas. The surgical removal of the thymus gland is used to induce a remission or decrease the symptoms of the disease. But questions remain: Do myasthenia gravis patients without thymomas benefit from the surgical removal of the thymus gland or would they be helped by medical treatment alone? Do the risks involved in thymectomy procedures justify their widespread use?

Trials to answer these questions are being conducted in almost 70 medical centers in 22 countries including the United States, Canada, South America, Europe, Australia, South Africa, and the Far East. Some of the leading American researchers in the thymectomy study are Drs. Richard Barohn, University of Kansas Medical Center; Janice Massey, Duke University Medical Center; Robert Pascuzzi, Indiana University School of Medicine; Shin Oh, University of Alabama; and Henry Kaminski, St. Louis University School of Medicine.

The MGFA has helped to devise the initial impetus and development plans for the study. Since 2000, the National Institute of Neurological Disorders and Stroke

(NINDS) has been sponsoring clinical trials comparing patients on prednisone treatment who have thymectomies with those who are on prednisone but do not have the surgery. Some of the specific answers these studies will investigate are:

- Does a combination of thymectomy and prednisone improve MG weakness *more than* prednisone alone?
- Does thymectomy plus prednisone enable myasthenics to decrease the dosage of prednisone? (If this proved to be true, it would reduce the risks of the drug.)
- Are there quality of life issues to consider in weighing the value of combining thymectomy and prednisone rather than prednisone alone?

If these clinical trials prove a direct link between using thymectomy surgery plus prednisone and decreasing the dosage of prednisone—reducing side effects, the value of combining these two treatments would be proven. On the other hand, if the combination of thymectomy and prednisone does not show the benefits of lowering drug dosage or decreasing side effects, then the opposite would be true.

The combined treatment plan would offer no advantage over giving prednisone alone. This topic is the focus of intensive clinical studies being conducted worldwide. The answers to this controversial dilemma in MG treatment need to be resolved.

THE IMMUNE SYSTEM—A CLOSER LOOK

In Chapter 9, we discussed MG as an autoimmune disorder. Antibodies produced by the immune system attack the neuromuscular junction causing weakness in muscle contraction. That basic explanation is only the beginning. If you look at MG research studies, you'll find additional terminology which requires a more in-depth understanding of the immune system. To help you fathom your way through this web of material, let's untangle some of the words you're likely to encounter.

Let's begin with unraveling T cells from B cells. T cells and B cells are two kinds of lymphocyte or white blood cells important in targeting immune responses. T cells are derived from or developed in the thymus gland; B cells are from the bone marrow. About 75% of all lymphocytes are T cells while 10% of lymphocytes are B cells and the rest have the ominous name of "killer cells." The B cells produce antibodies that target certain parts of proteins. But T cells can change or alter the activity of B cells almost like a traffic cop directs the flow of traffic on a roadway.

How does this fit into the myasthenia gravis picture? In MG, T cells modulate the activity of B cells which produce the antibodies (AChR) that attack acetylcholine receptor sites and other sites on the neuromuscular junction. For this reason, you will see myasthenia gravis referred to as a *T-cell driven, autoimmune disease.*

You may wonder why T cells can be so active since the thymus gland shrinks with age. At birth, the thymus gland weighs between 15 to 35 grams. It grows until puberty and then shrinks as its lymphoid tissue is replaced by fat tissue. An adult thymus generally contains about 5 grams of thymus tissue. Meanwhile, about 1% of the thymus cells become T cells which leave the thymus and migrate to the spleen, lymph nodes, and other lymph tissues. These are the T cells involved in immune responses.

It is also important to recognize that there are many antibodies involved in MG. Remember the antibody test described in Chapter 10? This accurate blood test confirms the diagnosis of MG in about 80% of patients with generalized myasthenia. These patients are positive for AChR antibodies. But there are different kinds of AChR antibodies—some of them are binding, others function as modulating, and still others are blocking antibodies.

Some of the MG patients negative for AChR antibodies have been found to have a distinctly different kind of antibodies. These antibodies attack muscle protein and are called MuSK (also known as muscle-specific receptor tyrosine kinase) antibodies. These distinctions in myasthenia gravis open up new and exciting areas of research. For example, will patients with MuSK antibodies react differently to drugs and therapies than the patients with AChR antibodies?

Dr. Donald Sanders at Duke University Medical Center recently completed a three-month study of 80 MG patients with MuSK antibodies. These patients were given a combination of CellCept and prednisone. It was hoped that the combination of prednisone and CellCept might be more beneficial and provide more of a boost in muscle strength than prednisone alone. Unfortunately, the combination did not prove to be advantageous for this group of MuSK positive myasthenics. But this is only a preliminary study in an exciting area of ongoing research.

Another study conducted by Aspreva Pharmaceuticals tested 176 MG patients. All of the patients were given prednisone plus either CellCept or a placebo. Once again, the results of prednisone alone rather than the combination of prednisone and CellCept proved to be significantly better. Is it possible that the prednisone masked the effects of the CellCept?

At the same time, a CellCept study group has been formed to design and coordinate future trials of this immunosuppressant. Perhaps future studies will analyze the advantage of CellCept without prednisone. Anecdotal data from a number of myasthenics who benefited from CellCept have been encouraging. More clinical trials are needed to support the general observations.

ISSUES THAT REMAIN

While myasthenia gravis has the distinction of being one of the most well understood autoimmune neuromuscular diseases, a number of vital issues remain. Dr. Robert Pascuzzi, Professor and Chairman of Neurology at the Indiana University School of

Medicine proposes the following vital questions still in need of answers:

Why does a person actually get myasthenia gravis? What initially triggers the autoimmune response? Is it genetically determined? Is it due to exposure to some type of infection or perhaps a toxic exposure? Is it a result of an initial malfunction in the thymus gland? Is it in fact a combination of factors that must occur in concert to trigger the disease?

Dr. Pascuzzi also lists the following issues that need to be addressed. Here is a shortened version of his major points:

1) Where does the disease come from?
2) How can it be prevented?
3) How can we detect MG earlier and improve sensitive and specific methods of diagnosis?
4) In the treatment of MG, how can we more effectively use drugs and therapies to reverse the damage to the neuromuscular junction so that it can function more efficiently?
5) Since the mechanism of MG is the best understood of all autoimmune diseases, how can MG research benefit the vast spectrum of other autoimmune disorders?

These matters of concern were compiled by Robert Pascuzzi, M.D. in the New Direction of the MGFA: The Research Initiative (2007).

As Chairman of the Medical Advisory Board for MGFA, Dr. Pascuzzi is on the forefront of establishing Centers of Excellence specializing in MG. The aim of these Centers is to provide quality standards for patient care, education, and research. They will also strengthen MGFA support chapters and stimulate the growth of new chapters. At the present time, this concept is in the proposal stage but it bodes well for the future of MG patient care and management.

ON THE HORIZON

In addition to the clinical studies described above, there are numerous other research projects being conducted around the world. I was fortunate to attend the 11th International Conference on Myasthenia Gravis and Related Disorders in May, 2007. Much of the following material was gleaned from that meeting of dedicated physicians and researchers.

One of the areas of intense research is testing the benefits of intravenous immune globulin (IVIg) in MG patients. Dr. Lorne Zinman of the Sunnybrook Health Sciences Centre in Toronto, Canada completed a study of 51 MG patients whose symptoms were worsening. Significant improvement was noted in patients treated with IVIg. The greatest progress was seen in patients with more severe weakness.

In a related study, Dr. Philippe Gajdos of the Hospital Raymond Poincare in Garches, France found that plasma exchange was more beneficial than IVIg for patients in MG crisis. Future studies comparing the efficacies of IVIg and plasma exchange are areas of interest.

An intriguing finding by a research group headed by Sara Fuchs, PhD, of the Weizmann Institute of Science, Rehovot, Israel, supports the idea that the beneficial portion of IVIg is a "disease-specific anti-immunoglobulin that is present in human IgG" (immunoglobulin G). If this proves to be true, it may be possible to remove this active ingredient from IVIg and discover a new and improved means of treating myasthenia gravis.

Other trials inhibiting complement (a group of 25 proteins in the blood that play various roles in the immune system) have proven effective in experimental rodents with autoimmune MG. These trials are likely to be extended to human studies in the future.

One of the most significant avenues of MG research is finding a way to eliminate the damaging AChR antibodies without upsetting the other functions of the immune system or causing serious side effects. Is there a way of switching off the T cells and B cells involved in MG antibody production? This type of targeted approach is the focus of studies conducted by Dr. Daniel B. Drachman of Johns Hopkins School of Medicine in Baltimore, MD. The results of these tests in the lab and animal models are encouraging. Applying this targeted approach in human studies is a goal for future work.

A drug called Monarsen, based on the research of Professor Hermona Soreq of Hebrew University in Israel, works quite differently than other MG medicines. Monarsen is classified as an antisense drug that penetrates the blood-brain barrier. It seems to work longer and more effectively than Mestinon, and has no known side effects. Clinical trials are underway in the UK and Israel.

At the 2007 MGFA meeting, Dr. Jon Sussman of the Greater Manchester Neuroscience Centre, UK, presented the dramatic results of testing Monarsen in patients withdrawn from anti-cholinesterase drugs. Patients taking Monarsen also exhibited fewer side effects, although the long-term benefits require further study.

Then there is the possibility of genetically altering muscle tissue. A certain defect in the genetic code for a substance called myostatin helps to build better and stronger muscles in animals. Could this same factor potentially apply to people with weak muscles?

Another exciting area of research is being conducted at the University of Alabama at Birmingham (UAB). Dr. J. Edwin Blalock and his associates at UAB are developing a therapeutic vaccine for MG. Unlike a vaccine that helps to prevent a disease from occurring, a therapeutic vaccine is designed to treat patients who already have the disease. The aim of this type of vaccine is to trigger a major reprieve or even a remission of symptoms. If it were possible to extend a respite from muscle weakness for a number of years, this would be a tremendous breakthrough.

Dr. Blalock's research into therapeutic vaccines is still in its early stages. Animal vaccine studies on rats with experimentally induced MG have proven to be effective. In dogs that had developed MG—without artificial means—the results are even more amazing.

According to Dr. Ronald E. Henderson in his book, *Attacking Myasthenia Gravis*, "...the use of this therapeutic vaccine has resulted in an 86% MG remission rate in the dogs that were treated."

Another group working on a therapeutic vaccine is led by Dr. Hideki Garren of Bayhill Therapeutics, Inc., Palo Alto, CA. They have begun clinical trials of a vaccine for multiple sclerosis and type 1 diabetes, and are now working on a vaccine for MG. So far they have successfully reversed symptoms in animal tests. Many more experiments and administrative hurdles are necessary before Food and Drug Administration approval for testing the vaccine in humans.

...So much research and hope for the future! A sincere thank you to all those who dedicate their lives to finding solutions.

FACTS IN A NUTSHELL

- Research scientists gather data to find solutions to the mystery of diseases. MG holds a unique position in that it is the most well understood of all the autoimmune, neuromuscular diseases.

- Many research facilities are conducting trials of the organ transplant drug CellCept, an immunosuppressant that may prove to be a safer alternative than prednisone.
- Thymectomy, the surgical removal of the thymus gland, has been a standard practice for MG patients. No distinction has been made between myasthenics with or without thymomas. A large scale study is now evaluating whether this practice is necessary for alleviating symptoms in MG patients without thymomas or if medical treatment would be sufficient.
- Questions regarding the effectiveness of combining thymectomy and prednisone are also being investigated, including the advantage of decreasing the dosage of prednisone in those individuals who have undergone thymus surgery.
- T cells and B cells are two kinds of lymphocytes or white blood cells involved in immune activity. T cells regulate the activity of B cells which produce AChR antibodies. Myasthenia gravis is a T-cell driven, autoimmune disease.
- There are different kinds of antibodies involved in MG. Various forms of AChR antibodies block, destroy or damage the neuromuscular junction. Another group of antibodies called MuSK (muscle-specific receptor tyrosine kinase) attack muscle protein.

- MG issues that need to be resolved include: what causes MG; how could it be prevented; are there more sensitive means of diagnosing the disease; and how drugs and therapies can be used to reverse damage to the neuromuscular junction.
- Some of the research projects on the horizon include testing the benefits of IVIg (intravenous immune globulin), inhibiting the complement component of the immune system, targeting the AChR antibodies without upsetting the immune system functioning, genetically altering muscle tissue, and developing a therapeutic vaccine.

My journey with myasthenia gravis continues with its twists and turns, highs and lows. I hope that my sharing helps you—the reader—in finding your own pathway. May God bless and strengthen you with His grace.

Made in the USA
Lexington, KY
26 May 2013